Frontiers in Mathematics

This series is designed to be a repository for up-to-date research results which have been prepared for a wider audience. Graduates and postgraduates as well as scientists will benefit from the latest developments at the research frontiers in mathematics and at the "frontiers" between mathematics and other fields like computer science, physics, biology, economics, finance, etc. All volumes are online available at SpringerLink.

More information about this series at http://www.springer.com/series/5388

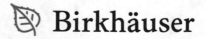

Ileana Bucur • Gavriil Paltineanu

Topics in Uniform Approximation of Continuous Functions

 Birkhäuser

Ileana Bucur
Department of Mathematics
and Computer Science
Technical University of Civil Engineering
Bucharest, Romania

Gavriil Paltineanu
Department of Mathematics
and Computer Science
Technical University of Civil Engineering
Bucharest, Romania

ISSN 1660-8046   ISSN 1660-8054   (electronic)
Frontiers in Mathematics
ISBN 978-3-030-48411-8   ISBN 978-3-030-48412-5   (eBook)
https://doi.org/10.1007/978-3-030-48412-5

This book is published under the imprint Birkhäuser, www.birkhauser-science.com, by the registered company
Springer Nature Switzerland AG.
The registered company address is: Gewerbestrasse 11, 6330 Cham, Switzerland

Introduction

The uniform approximation of continuous functions by simpler functions has its origins in Weierstrass' approximation theorems. Weierstrass published two results in 1885 when he was in his seventies where he shows that any continuous function on a real compact interval can be uniformly approximated by an algebraic polynomial, and, accordingly, any continuous periodic function of a period $T = 2\pi$ can be uniformly approximated by a trigonometric polynomial. These results somewhat counterbalance his earlier, famous example of a continuous function on a real interval, which is not differentiable at any point on that interval. This example from 1861 shows that some continuous functions can be "very non-smooth," while from Weierstrass approximation theorems it follows that a continuous function can be uniformly approximated by "very smooth" functions (i.e., polynomial functions).

It is well known that an entire function (a function that allows a power series expansion) can be uniformly approximated on any interval from the domain of convergence of its power series by an algebraic polynomial (the corresponding Taylor polynomial). However, it is also known that only a restricted class of functions can be expanded in a power series; such a function must be, among other requirements, infinitely differentiable. Weierstrass' theorem widely expands the class of functions that can be uniformly approximated by polynomials from entire functions to continuous functions.

Weierstrass theorems are so important—both from a theoretical and practical point of view—that ever since they were published numerous mathematicians proposed various proofs for them. These proofs are extremely instructive, as their basic ideas can be successfully applied in other problems of mathematical analysis.

In his excellent survey [32], Allan Pinkus classifies these proofs in three groups. The first group is composed of the proofs based on singular integrals; the original proof of Weierstrass, as well as the proofs given by Picard, Fejér, Landau and de la Vallée Poussin belong to this class. The second group contains the proofs based on the uniform approximation of some particular functions, such as the proofs given by Runge/Phragmén, Lebesgue, Mittag Leffler and Lerch. The proofs that do not belong to the first two groups form the third one; among these, we mention the proofs given by Bernstein, Volterra and Lerch.

All these proofs appeared prior to 1913. Half of a century later, in 1964, H. Kuhn published in Archiv der Mathematik an "elementary" proof of the theorem concerning the approximation by algebraic polynomials based on Bernoulli's inequality. Allan Pinkus considers this the simplest and the most elegant proof. Kuhn's proof is presented in Chap. 1, Sect. 1.1 of this book, along with the equivalence of the two Weierstrass' approximation theorems: by algebraic and by trigonometric polynomials. Kuhn's idea of using Bernoulli's inequality was imitated by other mathematicians to the purpose of giving new "elementary" proofs for other approximating theorems. In this respect, we must specify that a "more elementary proof" does not necessarily mean an "easier proof", but a proof using less "strong" mathematical results.

Chapter 1 also presents various generalizations of Weierstrass' theorems. Which leads to Sect. 1.2 in which we emphasize two elegant and consistent generalizations due to Korovkin and Bohman-Korovkin.

A very consistent generalization was obtained by M.H. Stone in 1937; it was re-published in 1948 with a simplified proof under the title "The Generalized Weierstrass Approximation Theorem," in the journal *Mathematics Magazine*. Stone's generalization follows two directions: on the one hand, passing from continuous functions on a closed real interval to functions which are continuous on an arbitrary compact space and, on the other hand, passing from the algebra of polynomials to a sub-algebra of $C(K, \mathbb{R})$ satisfying certain conditions. This result, presently known as the Stone-Weierstrass theorem, is presented in Sect. 1.3.

Like Weierstrass' theorem, the Stone-Weierstrass theorem has many proofs, more or less elementary, from the proof given by Louis de Branges in 1959, who uses results from functional analysis (Hahn-Banach and Krein–Milman's theorems), to the proof we present in Sect. 1.3 based on Brosowski and Deutsch's paper [5], where they use Bernoulli's inequality.

It is a well-known fact that Stone-Weierstrass' theorem is no more true in the case of complex functions, unless the subalgebra $\mathcal{A}$ of $C(K, \mathbb{C})$ is self-adjoint, i.e., $\forall\, a \in \mathcal{A}$ implies $\bar{a} \in \mathcal{A}$. Erret Bishop [2] obtained a generalization of Stone-Weierstrass' theorem to non-self-adjoint algebras. In Sect. 1.4. we present a very elegant proof of Bishop's theorem, due to T.J. Ransford [35], who also uses Bernoulli's inequality.

Chapter 1 ends with Sect. 1.5, entirely based upon the results obtained by Paltineanu and Bucur [31]. The first part of this section introduces the notion of Uryson family. A theorem of density in the space of continuous functions defined on a compact space with values in the interval [0, 1] is then proved; this is a key theorem from which one can also deduce Stone-Weierstrass' theorem, both in its algebraic and lattice variants as above. The second part of this section studies the subsets of $C(X, [0, 1])$ having the (VN) property (Von Neumann), and a Bishop type theorem for such sets is established. This theorem generalizes a result from 1992 due to Prolla the result which, in its turn, generalizes Von Neumann's variant of Stone-Weierstrass' theorem for continuous functions with range in [0,1]. We mention that both Prolla's proof and our proof use Bernoulli's inequality.

Following the generalization of the results from the first chapter, the second chapter presents various approximation theorems for continuous functions defined on a locally compact space. The non-compactness of the basic space gives rise to serious problems concerning the statement and the proof of such theorems. The problem of the kind of continuous functions we wish to deal with is taken into account: all of the continuous functions, only those which are continuous and bounded, those which vanish at infinity, those with compact support, etc. A natural frame for the study of such theorems proved to be that of the weighted spaces, introduced by Leopoldo Nachbin. A proper presentation of these spaces can be found in his paper [20].

The basic properties of the weighted spaces, as well as those of their duals, form the focus of Sects. 2.1 and 2.2. The second part of Chap. 2 presents some theorems of Stone-Weierstrass type for a vector subspace or a convex sub-cone of a weighted space. The chapter ends with the generalization of these results for vector functions.

In Chap. 3 we study the approximation of continuously differentiable functions. Bernstein's theorem and a Stone-Weierstrass type theorem, due to L. Nachbin (see e.g. [26, pp. 104, 107]), are presented.

Chapter 4, entirely based on the papers [14, 15, 21, 28–30], is devoted to the generalization of some of the previous results from the case of weighted spaces to the abstract case of the locally convex lattices of type (M). The real locally convex lattices of type (M) generalize the weighted spaces $CV_0(X, \mathbb{R})$, while the complex locally convex lattices of type (M) generalize the weighted spaces $CV_0(X, \mathbb{C})$. The idea of these generalizations belongs to Dan Vuza. It is generally known that a one-to-one correspondence exists between the closed subsets of a locally compact space X and the closed ideals of the weighted space $CV_0(X, \mathbb{R})$. More precisely,

$$S \xrightleftharpoons{\qquad} I_S = \{f \in CV_0(X, \mathbb{R}); \ f \,|S = 0\} \,.$$

This remark allows defining the notion of an antisymmetric ideal in a real locally convex lattice of type (M) by analogy with the notion of an antisymmetric set. Then two theorems of approximation for a vector subspace of a real locally convex lattice of type (M), in particular for a convex cone, are presented.

The complex vector lattices, less known in the literature, were nevertheless studied by numerous mathematicians, such as: Romulus Cristescu, H.H. Schaefer, H.P. Lotz, W.A. Luxemburg and A.C. Zaanen, G. Mittelmayer and M. Wolff, W.J. De Schipper [39], Dan Vuza, etc.

In Sect. 4.6, of preliminaries and notations, we used the paper.

Dan Vuza: *Elements of the theory of modules over ordered rings,* Order Structures in Functional Analysis, Vol **2**, Editura Academiei Române, Bucureşti, 1989, 175–283.

We conclude Chap. 4 by introducing the antisymmetric ideal in a complex locally convex lattice of type (M), by generalizing the de Branges lemma and obtaining a theorem of approximation of the elements of such a lattice by elements belonging to one of its vector subspaces.

Contents

Approximation of Continuous Functions on Compact Spaces

1.1 Weierstrass Approximation Theorems

The first Weierstrass theorem shows that the set of algebraic polynomials is uniformly dense in the space of continuous real functions on a compact interval of $\mathbb{R}$.

The second Weierstrass theorem asserts that the set of trigonometric polynomials is dense, with respect to the same topology, in the space of all real 2π-periodic, continuous functions on $\mathbb{R}$.

In this section, we present the proof, given by H. Khun, of the first Weierstrass approximation theorem, followed by a proof of the equivalence of the above two assertions.

Theorem 1.1.1 *Let k be a natural number, $k > 1$, let $a, b \in (0, 1)$ be such that $a < \frac{1}{k} < b$ and, for any natural number, let $p_n : [0, 1] \to [0, 1]$ be the function given by*

$$p_n(t) = \left(1 - t^n\right)^{k^n}, \ \forall t \in [0, 1].$$

Then the sequence $(p_n)_n$ of polynomial functions converges uniformly to the constant function 1 on the interval $[0, a]$ (or 0 on $[b, 1]$).

Proof By Bernoulli inequality, we get

$$t \in [0, a] \Rightarrow 1 - (ak)^n \le 1 - (tk)^n \le p_n(t) = (1 - t^n)^{k^n} \le 1,$$

$$t \in [b, 1] \Rightarrow 0 \le p_n \le \frac{1}{(1 + t^n)^{k^n}} \le \frac{1}{1 + (tk)^n} \le \frac{1}{1 + (bk)^n} \le \frac{1}{(bk)^n}.$$

I. Bucur, G. Paltineanu, *Topics in Uniform Approximation of Continuous Functions*, Frontiers in Mathematics, https://doi.org/10.1007/978-3-030-48412-5_1

Since $ak<1<bk$, we deduce that the sequences $\left((a \cdot k)^n\right)_n$ and $\left(\left(\frac{1}{b \cdot k}\right)^n\right)_n$ are convergent to 0 and therefore the stated assertions are proved.   □

Corollary 1.1.2 (H. Khun, see e.g. [32, p. 7]) *For any* $\delta \in (0, 1)$ *the sequence* $(p_n)_n$ *of polynomial functions,* $p_n : [-1, 1] \to [0, 1]$, *given by*

$$p_n(x) = \left(1 - \left(\frac{1-x}{2}\right)^n\right)^{2^n}$$

is uniformly convergent to 0 *(resp. to* 1*) on the interval* $[-1, \delta]$ *(resp.* $[\delta, 1]$*).*

Proof The function $u : [-1, 1] \to [0, 1]$ given by

$$u = u(x) = \frac{1-x}{2}, \ \forall x \in [-1, 1]$$

is a strictly decreasing bijection and we have

$$0 = u(1) < \frac{1-\delta}{2} = u(\delta) < \frac{1}{2} = u(0) < \frac{1+\delta}{2} = u(-\delta) < u(-1) = 1.$$

Taking now $k = 2, a = \frac{1-\delta}{2}, b = \frac{1+\delta}{2}$ in Theorem 1.1.1, we get that the sequence: $q_n : [0, 1] \to [0, 1]$, given by

$$q_n(u) = \left(1 - u^n\right)^{2^n}, \ \forall u \in [0, 1],$$

converges uniformly to 1 (resp. to 0) on the interval $\left[0, \frac{1-\delta}{2}\right]$ (resp. $\left[\frac{1+\delta}{2}, 1\right]$). The proof is finished if we remark that $p_n(x) = q_n[u(x)]$ and $u\left([-1, \delta]\right) = \left[\frac{1+\delta}{2}, 1\right]$ and

$$u\left([\delta, 1]\right) = \left[0, \frac{1-\delta}{2}\right].$$ □

Theorem 1.1.3 (Weierstrass [43]) *Let* $f : [0, 1] \to \mathbb{R}$ *be a continuous function. Then for any* $\forall \varepsilon > 0$ *there exists a polynomial function* P_ε *such that*

$$|f(x) - P_\varepsilon(x)| < \varepsilon, \ \forall x \in [0, 1].$$

Proof First of all, we replace the function f by a spline function g of first order such that

$$|f(x) - g(x)| < \frac{\varepsilon}{2}, \ \forall x \in [0, 1].$$

By uniform continuity property of f, we may consider a system $(x_i)_{1 \le i \le n}$ of real numbers:

$$0 = x_0 < x_1 < \ldots < x_{i-1} < x_i < \ldots < x_n = 1$$

such that the oscillation ω_i of the function f on the interval $[x_{i-1}, x_i]$ is smaller than $\frac{\varepsilon}{2}$, i.e., $M_i - m_i < \frac{\varepsilon}{2}$, where $M_i = \sup\{f(t);\ t \in [x_{i-1}, x_i]\}, m_i = \inf\{f(t);\ t \in [x_{i-1}, x_i]\}$. Then we consider the affine functions $g_i : [0, 1] \to \mathbb{R}$, given by

$$g_i(x) = f(x_{i-1}) + \frac{f(x_i) - f(x_{i-1})}{x_i - x_{i-1}}(x - x_{i-1}),$$

and we remark that $g_i(x_{i-1}) = f(x_{i-1})$, $g_i(x_i) = f(x_i)$, hence $g_i(x_{i-1})$, $g_i(x_i)$ belong to the interval $[m_i, M_i]$, and therefore for any $x \in [x_{i-1}, x_i]$, we have

$$g_i(x),\ f(x) \in [m_i, M_i],\ |g_i(x) - f(x)| < \frac{\varepsilon}{2}.$$

Further we denote g the function defined on the interval $[0, 1]$ by

$$g(x) = g_1(x) + \sum_{i=1}^{n-1} [g_{i+1}(x) - g_i(x)] \cdot h(x - x_i),$$

where h is the Heaviside function on R:

$$h(t) = \begin{cases} 0,\ if\ t < 0 \\ 1,\ if\ t \ge 0. \end{cases}$$

It is easy to verify that for any $i \in \{1, 2, \ldots, n\}$ the function g_i is the restriction of the function g to $[x_{i-1}, x_i]$ and therefore $|f(x) - g(x)| < \frac{\varepsilon}{2}, \forall x \in [0, 1]$.

We remark now that for any $i \in \{1, 2, \ldots, n - 1\}$ we have

$$g_{i+1}(x_i) - g_i(x_i) = f(x_i) - f(x_i) = 0,$$

and therefore, by continuity, we have $|g_{i+1}(x) - g_i(x)| < \frac{\varepsilon}{2 \cdot n}$ if $|x - x_i| < \delta$ for a suitable $\delta > 0$.

Since $x - x_i \in [-1, 1]$ for all $x \in [0, 1]$ and any $i \in \{1, 2, \ldots, n\}$, using the notations from the previous corollary and noting

$$\Delta_n(x) = g(x) - \left(g_1(x) + \sum_{i=1}^{n-1} [g_{i+1}(x) - g_i(x)] \cdot p_n(x - x_i) \right)$$

$$= \sum_{i=1}^{n-1} [g_{i+1}(x) - g_i(x)] \cdot [h(x - x_i) - p_n(x - x_i)]$$

we have

$$|\Delta_n(x)| < \frac{\varepsilon}{2},$$

if $|x - x_i| > \delta$ and $n \in \mathbb{N}$ is sufficiently large. For $|x - x_i| < \delta$ it results

$$|\Delta_n(x)| \le \sum_{i=1}^{n-1} |g_{i+1}(x) - g_i(x)| < n \cdot \frac{\varepsilon}{2 \cdot n} = \frac{\varepsilon}{2},$$

and therefore

$$\left| f(x) - \left(g_1(x) + \sum_{i=1}^{n-1} [g_{i+1}(x) - g_i(x)] \cdot p_n(x - x_i)]\right) \right| \le |f(x) - g(x)| + |\Delta_n(x)|$$

$$< \varepsilon, \forall x \in [0, 1].$$

As the function $P_\varepsilon(x) = g_1(x) + \sum_{i=1}^{n-1} [g_{i+1}(x) - g_i(x)] \cdot p_n(x - x_i)]$ is a polynomial function, the proof is finished.   □

Remark 1.1.4   The Weierstrass theorem may be stated under the following form: Any real continuous function on the interval $[a, b]$, $a < b$ is the uniform limit of a sequence $(P_n)_n$ of polynomial functions on $[a, b]$.

Indeed, the affine function $\varphi : [0, 1] \to [a, b]$ given by

$$\varphi(t) = a + t \cdot (b - a), t \in [0, 1]$$

is a homeomorphism having as inverse the map ψ given by

$$\psi(x) = \frac{x - a}{b - a}, \forall x \in [a, b].$$

If f is a real continuous function on $[a, b]$, then the function $f \circ \varphi$ defined on $[0, 1]$ is continuous, and therefore there exists a sequence $(p_n)_n$ of polynomial functions which converges uniformly to the function $f \circ \varphi$ on $[0, 1]$. Hence the sequence $(P_n)_n$ of polynomial functions on $[a, b]$, given by $P_n = p_n \circ \psi$, converges uniformly to the function $f = f \circ \varphi \circ \psi$.

We recall that a trigonometric polynomial is a function $t : \mathbb{R} \to \mathbb{R}$ of the form

$$t(x) = \sum_{k=0}^{m} (a_k \cdot \cos k \cdot x + b_k \cdot \sin k \cdot x).$$

Remark 1.1.5 For any polynomial function $p(x) = \sum_{i=1}^{m} c_i \cdot x^i$, $x \in \mathbb{R}$, the functions

$$x \to p(\sin x), \ x \to p(\cos x)$$

are trigonometric polynomials.

We may express the functions $x \to \cos^n x$, $x \to \sin^n x$ as trigonometric polynomials using the well-known identities:

$$\cos^2 x = \frac{1 + \cos 2x}{2}, \ \sin^2 x = \frac{1 - \cos 2x}{2}, \ \cos x \cdot \cos y = \frac{1}{2}[\cos(x+y) + \cos(x-y)],$$

$$\sin x \cdot \sin y = \frac{1}{2}[\cos(x-y) - \cos(x+y)], \ \sin x \cdot \cos y = \frac{1}{2}[\sin(x+y) + \sin(x-y)].$$

Theorem 1.1.6 (Weierstrass [43]) *Let $f : \mathbb{R} \to \mathbb{R}$ be a continuous 2π-periodic function. Then there exists a sequence $(T_n)_n$ of trigonometric polynomial which converges uniformly on $\mathbb{R}$ to f.*

Proof First we suppose that the function f is also symmetric, i.e., $f(-t) = f(t)$ for all $t \in \mathbb{R}$. Further we consider the continuous function $\varphi : [-1, 1] \to \mathbb{R}$ given by

$$\varphi(x) = f(\arccos x),$$

and the polynomial sequence of functions $(P_n)_n$ such that P_n is uniformly convergent to φ on the set $[-1, 1]$. Obviously the sequence of trigonometric polynomials $(T_n)_n$, $T_n = P_n \circ \cos$ converges uniformly to the function f on the interval $[0, \pi]$. But the trigonometric polynomials P_n are all symmetric and so is the function f. Hence the sequence $(T_n)_n$ converges uniformly to f on $[-\pi, \pi]$, and therefore on $\mathbb{R}$, by periodicity.

Generally, we show that the function $S : \mathbb{R} \to \mathbb{R}$, defined by

$$S(t) = f(t) \cdot \sin^2 t, \ t \in \mathbb{R},$$

is uniformly approximated by trigonometric polynomials on $\mathbb{R}$.

Indeed, using the above considerations we may consider two sequences $(T_{1n})_n$, $(T_{2n})_n$ of trigonometric polynomials which are uniformly convergent on $\mathbb{R}$ to the symmetric functions g, h given by

$$g(t) = \frac{1}{2}(f(t) + f(-t)), \ h(t) = \frac{1}{2}(f(t) - f(-t)) \cdot \sin t.$$

We have

$$f(t) \cdot \sin^2 t = g(t) \cdot \sin^2 t + h(t) \cdot \sin t,$$

and therefore the trigonometric sequence of polynomials $(T_n)_n$ defined by

$$T_n(t) = \sin^2 t \cdot T_{1n}(t) + \sin t \cdot T_{2n}(t),$$

converges uniformly on $\mathbb{R}$ to the function S.

We consider now a sequence of trigonometric polynomials $(Q_n)_n$ which is uniformly convergent on $\mathbb{R}$ to the function $C : \mathbb{R} \to \mathbb{R}$, given by

$$C(t) = f\left(\frac{\pi}{2} - t\right) \cdot \sin^2 t.$$

Obviously the sequence of trigonometric polynomials $(R_n)_n$, given by

$$R_n(t) = Q_n\left(\frac{\pi}{2} - t\right)$$

is uniformly convergent on R to the function

$$t \to C\left(\frac{\pi}{2} - t\right) = f(t) \cdot \sin^2\left(\frac{\pi}{2} - t\right) = f(t) \cdot \cos^2 t.$$

Hence the sequence $(T_n + R_n)_n$ is uniformly convergent on $\mathbb{R}$ to the function f.   □

Remark 1.1.7 The above theorem shows how the approximation theorem of continuous functions on compact intervals by algebraic polynomials implies the uniform approximation of 2π-periodic continuous functions on $\mathbb{R}$ by trigonometric polynomials. The converse implication is also true.

Indeed, let $f : [0, 1] \to \mathbb{R}$ be a continuous function and let $\tilde{f}$ be a continuous extension of f to the interval $[-\pi, \pi]$ such that $\tilde{f}(-\pi) = \tilde{f}(\pi) = 0$. It is easy to verify that

$$\tilde{f}(t) = \begin{cases} \frac{f(0)}{\pi} \cdot (x + \pi), \, if \, x \in [-\pi, 0) \\ f(x) \quad, \, if \, x \in [0, 1] \\ \frac{f(1)}{1-\pi} \cdot (x - \pi), \, if \, x \in (1, \pi]. \end{cases}$$

Obviously, $\tilde{f}$ may be extended by periodicity to $\mathbb{R}$, and for any $\varepsilon > 0$, we may consider a trigonometric polynomial P, $P(t) = \sum_{k=1}^{m} (a_k \cdot \cos k \cdot t + b_k \cdot \sin k \cdot t)$ such that

$$\left|\tilde{f}(t) - P(t)\right| < \frac{\varepsilon}{2}, \forall t \in \mathrm{R}.$$

Using now Taylor uniform approximation of each function $t \to \cos k \cdot t$, $t \to \sin k \cdot t$, $k = 1, 2, \ldots, m$ by polynomial functions on the compact interval $[-\pi, \pi]$, we may choose the algebraic polynomials C_k, S_k such that

$$|a_k \cdot \cos k \cdot t - C_k(t)| < \frac{\varepsilon}{2 \cdot m}, \ |b_k \cdot \sin k \cdot t - S_k(t)| < \frac{\varepsilon}{2 \cdot m}$$

for all $k = 1, 2, \ldots, m$. Obviously, for any $t \in [-\pi, \pi]$ we have

$$\left| \tilde{f}(t) - \sum_{k=0}^{m} (S_k + C_k)(t) \right| < \left| \tilde{f}(t) - P(t) \right| + \left| P(t) - \sum_{k=0}^{m} (S_k + C_k)(t) \right| < \varepsilon.$$

1.2 Korovkin Type Theorems

In this section we denote by $C([0, 1])$ the set of all real continuous functions defined on the interval $[0, 1]$.

Definition 1.2.1 A linear operator $T : C([0, 1]) \to C([0, 1])$ will be called positive if $T(f) \geq 0$ for any positive function $f \in C([0, 1])$.

Further we endow the real vector space $C([0, 1])$ with the uniform topology given by the norm:

$$\|f\| = \sup \{|f(t)| ; t \in [0, 1]\}.$$

The following assertion is obvious.

Proposition 1.2.2 *Any positive linear operator on $C([0, 1])$ is continuous.*

Indeed, let $T : C([0, 1]) \to C([0, 1])$ be a linear and positive operator. Then we have

$$-|f| \leq f \leq |f| \ \Rightarrow \ -T(|f|) \leq T(f) \leq T(|f|)$$

and further

$$|T(f)| \leq T(|f|). \tag{1.1}$$

On the other hand, from the inequality $|f| \leq 1 \cdot \|f\|$ it results:

$$T(|f|) \leq T(1) \cdot \|f\|. \tag{1.2}$$

From the inequalities (1.1) and (1.2) it follows:

$$|T(f)| \leq T(1) \cdot \|f\|.$$

Theorem 1.2.3 (Korovkin [13]) *Let* p_0, p_1, p_2 *be the polynomial functions on* $[0, 1]$ *defined by*

$$p_0(x) = 1, \ \ p_1(x) = x, \ \ p_2(x) = x^2, \ \forall x \in [0, 1],$$

and let $(L_n)_n$ *be a sequence of positive, linear operators on* $C([0, 1])$ *such that*

$$\lim_{n \to \infty} \|L_n(p_i) - p_i\| = 0, i = 1, 2, 3.$$

Then, for any $f \in C([0, 1])$ *the sequence* $(L_n(f))_n$ *of functions of* $C([0, 1])$ *converges uniformly to the function* f, *i.e.,*

$$\lim_{n \to \infty} \|L_n(f) - f\| = 0.$$

Proof Using the uniform continuity of the function f on the interval $[0, 1]$, for any $k \in \mathbb{N}^*$ we choose $\delta_k > 0$ such that $\left| f(x') - f(x'') \right| < \frac{1}{k}$ if $\left| x' - x'' \right| < \delta_k$ and then we consider, for any $k \neq 0$ and $t \in [0, 1]$ the polynomial functions of second degree $q_{k,t}^-$, $q_{k,t}^+$ defined by

$$q_{k,t}^-(t) = -\frac{1}{k} + f(t) - M \cdot \frac{(x - t)^2}{\delta_k^2}, \ \ q_{k,t}^+(t) = \frac{1}{k} + f(t) + M \cdot \frac{(x - t)^2}{\delta_k^2}$$

where $M = 2 \cdot \|f\|$. One can see that

$$q_{k,t}^-(x) \leq f(x) \leq q_{k,t}^+(x), \forall x \in [0, 1].$$

Indeed, if $|t - x| < \delta_k$, then we have

$$f(x) - q_{k,t}^+(x) = f(x) - f(t) - \frac{1}{k} - 2 \cdot \|f\| \cdot \frac{(x - t)^2}{\delta_k^2} < f(x) - f(t) - \frac{1}{k} < \frac{1}{k} - \frac{1}{k} = 0.$$

If $|t - x| \geq \delta_k$, then

$$f(x) - q^+{}_{k,t}(x) = f(x) - f(t) - \frac{1}{k} - 2 \cdot \|f\| \cdot \frac{(x - t)^2}{\delta_k^2} < f(x) - f(t) - \frac{1}{k} - 2 \|f\| < -\frac{1}{k} < 0.$$

Therefore we have

$$f(x) < q^+_{k,t}(x), \ \forall x \in [0, 1], \ \forall t \in [0, 1].$$

Similarly it shows that

$$q^-_{k,t}(x) < f(x), \ \forall x \in [0, 1], \ \forall t \in [0, 1].$$

Using now the positivity of L_n, we deduce the inequalities:

$$L_n\left(q^-_{k,t}\right) \leq L_n(f) \leq L_n\left(q^+_{k,t}\right)$$

$$-q^+_{k,t} \leq -f \leq -q^-_{k,t}.$$

Using the notation $M_k = \frac{M}{\delta_k{}^2}$ and adding the previous inequalities, we obtain

$$\left(-\frac{1}{k} + f(t)\right) \cdot L_n(p_0) - M_k \cdot L_n\left((p_1 - t \cdot p_0)^2\right) - \frac{1}{k} - f(t) - M_k \cdot \left((p_1 - t \cdot p_0)^2\right) \leq L_n(f) - f$$

$$\leq \left(\frac{1}{k} + f(t)\right) \cdot L_n(p_0) + M_k \cdot L_n\left((p_1 - t \cdot p_0)^2\right) + \frac{1}{k} - f(t) + M_k \cdot \left((p_1 - t \cdot p_0)^2\right) \quad (1.3)$$

Now, by hypotheses we have

$$\lim_{n \to \infty} M_k \cdot \left\| L_n(p_1{}^2) - p_1{}^2 \right\| = \lim_{n \to \infty} M_k \cdot t \cdot \| L_n(p_1) - p_1 \| = \lim_{n \to \infty} M_k \cdot t^2 \cdot \| L_n(p_0) - p_0 \| = 0.$$

and therefore

$$\lim_{n \to \infty} \left\| M_k \cdot L_n((p_1 - t \cdot p_0)^2) - M_k \cdot (p_1 - t \cdot p_0)^2 \right\| = 0$$
and
$$\lim_{n \to \infty} |f(t)| \cdot \| L_n(p_0) - p_0 \| \leq \lim_{n \to \infty} M \cdot \| L_n(p_0) - p_0 \| = 0.$$

Let now $y \in [0, 1]$ be an arbitrary point. From the preceding considerations and the inequality (1.3), we get

$$-\frac{1}{k} \cdot (L_n(p_0) - p_0)(y) + f(t) \cdot (L_n(p_0) - p_0)(y) - M_k$$

$$\cdot L_n\left((p_1 - t \cdot p_0)^2\right)(y) - M_k \cdot (p_1 - t \cdot p_0)^2(y)$$

$$\leq L_n(f)(y) - f(y) \leq \frac{1}{k} \cdot (L_n(p_0) - p_0)(y) + f(t) \cdot (L_n(p_0) - p_0)(y)$$

$$+ M_k \cdot L_n \left((p_1 - t \cdot p_0)^2 \right)(y) + M_k \cdot (p_1 - t \cdot p_0)^2(y).$$

On the other hand, we have

$$\left| \frac{1}{k} \cdot (L_n(p_0) - p_0)(y) \right| \leq \frac{1}{k} \cdot \| L_n(p_0) - p_0 \|, \ \forall k \in \mathbb{N}^*, \ \forall n \in \mathbb{N}^*$$

$$|f(t) \cdot (L_n(p_0) - p_0)(y)| \leq M \cdot \| L_n(p_0) - p_0 \|, \forall t \in [0, 1], \forall n \in \mathbb{N}^*$$

and

$$\left| M_k \cdot L_n (p_1 - t \cdot p_0)^2 (y) \right| \leq \left\| M_k \cdot L_n (p_1 - t \cdot p_0)^2 - M_k \cdot (p_1 - t \cdot p_0)^2 \right\|$$

$$+ \left| M_k \cdot (p_1 - t \cdot p_0)^2 (y)) \right|, \ \ \forall n, k \in \mathbb{N}^*.$$

We remark that, taking $t = y$ then $M_k \cdot (p_1 - t \cdot p_0)(y) = 0$ and for any $\varepsilon > 0$, we can fix k sufficiently large such that

$$\frac{1}{k} \cdot \| L_n(p_0) - p_0 \| \leq \frac{\varepsilon}{4}, \ \text{for all } n \in \mathbb{N}^*.$$

And then, we take $n_\varepsilon \in \mathbb{N}^*$ such that for all $n \geq n_\varepsilon$, $n \in \mathbb{N}^*$, we have

$$M \cdot \| L_n(p_0) - p_0 \| \leq \frac{\varepsilon}{4}, \ \ \left\| M_k \cdot L_n \left((p_0 - t \cdot p_0)^2 \right) - M_k \cdot (p_1 - t \cdot p_0)^2 \right\| < \frac{\varepsilon}{4},$$

i.e., $|(L_n(f) - f)(y)| < \varepsilon$, for all $n \in \mathbb{N}^*$ and all $y \in [0, 1]$. □

Definition 1.2.4 The linear, positive operator B_n, $n \in \mathbb{N}^*$, $B_n : C([0, 1]) \to C([0, 1])$ given by

$$B_n(f)(x) = \sum_{k=0}^{n} f\left(\frac{k}{n} \right) \cdot C_n{}^k \cdot x^k \cdot (1 - x)^{n-k}, \forall f \in C([0, 1])$$

will be called Bernstein operator of order.

Corollary 1.2.5 (Bernstein, see e.g. [32, p. 30]) *For any element* $f \in C([0, 1])$ *the sequence* $(B_n(f))_n$ *is uniformly convergent to* f *on the interval* $[0, 1]$.

Since $B_n(f)$ is a polynomial function, we get a new proof of the first Weierstrass theorem.

Proof We shall use the previous Korovkin theorem. It is sufficient to show that the sequences $(B_n(p_i))_n$, $i = 0, 1, 2$ are uniformly convergent to p_0, p_1, p_2, respectively. (We use the notations from Theorem 1.2.3.) We have

$$B_n(p_0) = \sum_{k=0}^{n} C_n^k \cdot x^k \cdot (1 - x)^{n-k} = [x + (1 - x)]^n = 1 = p_0(x),$$

hence

$$\|B_n(p_0) - p_0\| = 0, \ \forall n \in \mathbb{N}^*.$$

Let us denote $p_{nk}(x) = C_n^k \cdot x^k \cdot (1 - x)^{n-k}$, $\forall k \leq n$. Further we have

$$\sum_{k=0}^{n} k \cdot p_{nk}(x) = nx \cdot \sum_{k=1}^{n} \frac{(n-1)!}{(k-1)!(n-k)!} \cdot x^{k-1} \cdot (1 - x)^{n-k}$$

$$= nx \cdot \sum_{l=0}^{n-1} C_{n-1}^l \cdot x^l \cdot (1 - x)^{n-1-l} = nx,$$

and therefore

$$B_n(p_1)(x) = \sum_{k=0}^{n} \frac{k}{n} \cdot p_{nk}(x) = x = p_1(x), \|B_n(p_1) - p_1\| = 0, \ \forall n \in \mathbb{N}^*.$$

Analogously we have

$$\sum_{k=0}^{n} k \cdot (k - 1) \cdot p_{nk}(x)$$

$$= n \cdot (n - 1) \cdot x^2 \cdot \sum_{k=2}^{n} \frac{(n-2)!}{(k-2)!(n-k)!} \cdot x^{k-2} \cdot (1 - x)^{n-k}$$

$$= n \cdot (n - 1) \cdot x^2 \cdot \sum_{l=0}^{n-2} C_{n-2}^l \cdot x^l \cdot (1 - x)^{n-2-l} = n \cdot (n - 1)x^2$$

and

$$\sum_{k=0}^{n} k^2 \cdot p_{nk}(x) = n \cdot (n-1) \cdot x^2 + n \cdot x; \ B_n(p_2)(x)$$

$$= \frac{1}{n^2} \cdot \sum_{k=0}^{n} k^2 \cdot p_{nk}(x) = \frac{n \cdot (n-1)}{n^2} \cdot x^2 + \frac{x}{n}.$$

Therefore we have

$$\|B_n(p_2) - p_2\| = \sup_{x \in [0,1]} \left| \frac{n(n-1)x^2 + nx}{n^2} - x^2 \right| = \sup_{x \in [0,1]} \frac{|x - x^2|}{n} \le \frac{1}{n},$$

from which it follows that

$$\lim_{n \to \infty} \|B_n(p_2) - p_2\| = 0.$$

So, $\lim_{n \to \infty} \|B_n(p_i) - p_i\| = 0$, $i = 0, 1, 2$, and the Korovkin theorem may be used.

Further we present Bohman–Korovkin theorem, a strong generalization of the above Korovkin theorem.

For this we start with a Hausdorff compact space K, with the set $C(K)$ of all real continuous functions on K endowed with the topology of uniform convergence. For any $f \in C(K)$ we denote as usually by $\|f\|$ the uniform norm of f.   □

Theorem 1.2.6 (Bohman–Korovkin, see e.g. [34, p. 324]) *Let us suppose that K contains at least two points and let $L_n : C(K) \to C(K)$, $k \in \mathbb{N}^*$, be a sequence of positive linear operators. Let $f_i \in C(K)$, $i \in \overline{1, m}$ have the following properties:*

(i) $(L_n(f_i))_n$ is uniformly convergent to f_i for all $i \in \overline{1, m}$.
(ii) There exists $a_i \in C(K)$, $i = \overline{1, m}$ such that

$$\sum_{i=0}^{m} a_i(y) \cdot f_i(x) \ge 0, \ \forall (x, y) \in K \times K \text{ and } \sum_{i=0}^{m} a_i(y) \cdot f_i(x) = 0, \text{ iff } x = y.$$

Then, for any $f \in C(K)$ the sequence $(L_n(f))_n$ is uniformly convergent to f.

Proof For any $y \in K$ we denote $P_y = \sum_{i=1}^{m} a_i(y) \cdot f_i$ and we remark that for any real number ε, $\varepsilon > 0$ there exists $n_\varepsilon \in \mathbb{N}$ such that

$$\|L_n(P_y) - P_y\| \le \varepsilon, \ \forall n \ge n_\varepsilon, \ \forall y \in K.$$

Indeed, the functions a_i, $i \in \overline{1, m}$ are uniformly bounded on K, i.e., there exists $M \in \mathbb{R}_+^*$ such that $\|a_i\| \le M$, $\forall i \in \overline{1, m}$. Further we have

$$\left\| L_n(P_y) - P_y \right\| \le \sum_{i=1}^{m} \| a_i(y) \cdot (L_n(f_i) - f_i) \| \le M \cdot \sum_{i=1}^{m} \| L_n(f_i) - f_i \|,$$

and the last term is smaller then ε, if we choose $n_\varepsilon \in \mathbb{N}^*$ such that $\| L_n(f_i) - f_i \| < \frac{\varepsilon}{m \cdot M}$ for all $n \ge n_\varepsilon$ and all $i \in \overline{1, m}$. Particularly, the sequence of continuous functions on K:

$$y \to L_n\left(P_y\right)(y) : K \to \mathbb{R}$$

is uniformly convergent to 0 on K. Indeed, we have

$$\left| L_n\left(P_y\right)(y) \right| = \left| L_n\left(P_y\right)(y) - P_y(y) \right| \le \left\| L_n\left(P_y\right) - P_y \right\| < \varepsilon, \ \forall n \ge n_\varepsilon.$$

We consider now two points $y_1, y_2 \in K$, $y_1 \neq y_2$ and we denote by Q the element of $C(K)$ given by

$$Q = P_{y_1} + P_{y_2}.$$

Since $P_{y_1}(x) = 0$ if $x = y_1$ and $P_{y_2}(x) = 0$ if $x = y_2$, we deduce that Q is a strictly positive function on K and

$$\lim_{n \to \infty} \| L_n(Q) - Q \| = 0.$$

Let f be an arbitrary element of $C(K)$ and let

$$(y, x) \to g_y(x) = f(x) - \frac{f(y)}{Q(y)} \cdot Q(x) : K \times K \to \mathbb{R}$$

be a continuous function on $K \times K$ associated with f. Since this function vanishes on the diagonal Δ of $K \times K$, we deduce that for any $m \in \mathbb{N}^*$, the set U_m of $K \times K$,

$$U_m = \left\{ (y, x) \in K \times K; \ |g_y(x)| < \frac{1}{m} \right\}$$

is an open neighbourhood of Δ. On the compact subset $(K \times K) \backslash U_m$ the above associated function g_y is bounded, and the continuous function $(y, x) \to P_y(x)$ is strictly positive.

Let us denote

$$\alpha_m = \inf \left\{ P_y(x); \ (y, x) \in (K \times K) \backslash U_m \right\}$$

and

$$\beta_m = \sup \left\{ \left| g_y(x) \right| ; \ (y, x) \in (K \times K) \backslash U_m \right\}.$$

Just from the definitions of U_m, α_m, β_m, we have

$$\left| g_y(x) \right| \le \frac{1}{m} + \frac{\beta_m}{\alpha_m} \cdot P_y(x) , \ \forall (y, x) \in K \times K.$$

Since L_n is linear and positive, we get

$$\left| L_n(g_y) \right| \le \frac{1}{m} \cdot L_n(1) + \frac{\beta_m}{\alpha_m} \cdot L_n(P_y) , \ \text{on } K.$$

On the other hand the sequence $(L_n(Q))_n$ is uniformly convergent to Q on K and so, Q being strictly positive, the sequence $(L_n(1))_n$ is uniformly bounded, i.e., $\| L_n (1.1) \| \le M'$, for all $n \in \mathbb{N}$.

Hence for all $n \in \mathbb{N}, \ y \in K$ we have

$$\left| L_n(g_y)(y) \right| \le \frac{1}{m} \cdot M' + \frac{\beta_m}{\alpha_m} \cdot L_n(P_y)(y),$$

$$\left| L_n(f)(y) - \frac{f(y)}{Q(y)} \cdot L_n(Q)(y) \right| \le \frac{1}{m} \cdot M' + \frac{\beta_m}{\alpha_m} \cdot L_n(P_y)(y),$$

$$\left| L_n(f)(y) - f(y) \right| \le \left| L_n(f)(y) - \frac{f(y)}{Q(y)} \cdot L_n(Q)(y) \right| + \left| f(y) \left(1 - \frac{L_n(Q)(y)}{Q(y)} \right) \right|$$

$$\le \frac{1}{m} \cdot M' + \frac{\beta_m}{\alpha_m} \cdot L_n(P_y)(y) + \| f \| \cdot \left| \frac{L_n(Q)(y)}{Q(y)} - 1 \right|.$$

Since $Q > 0$ and the sequence $(L_n(Q))_n$ is uniformly convergent to Q we deduce that

$$\lim_{n \to \infty} \left\| \frac{L_n(Q)}{Q} - 1 \right\| = 0.$$

From the starting considerations the sequence, of functions $y \to L_n(P_y)(y)$ is uniformly convergent to 0. The number M' is independent of $m \in \mathbb{N}$. The numbers α_m, β_m depend on m, but we fix $\varepsilon > 0$ and we choose a m sufficiently large such that $\frac{M'}{m} < \frac{\varepsilon}{3}$. Then we take $n_\varepsilon \in \mathbb{N}^*$ such that

$$\left| \frac{\beta_m}{\alpha_m} \cdot L_n(P_y)(y) \right| < \frac{\varepsilon}{3} \ \text{for all } y \in K , \ \| f \| \cdot \left\| \frac{L_n(Q)}{Q} - 1 \right\| < \frac{\varepsilon}{3}, \forall n \ge n_\varepsilon,$$

and in this way

$$|L_n(f)(y) - f(y)| < \varepsilon, \ \forall y \in K, \text{ if } n \geq n_\varepsilon. \qquad \square$$

The following corollary shows that the Bohman–Korovkin theorem generalizes the old Korovkin theorem.

Corollary 1.2.7 (Korovkin [13]) *Let* p_0, p_1, p_2 *be the polynomial functions on* $[0, 1]$ *defined by*

$$p_0(x) = 1, \ p_1(x) = x, \ p_2(x) = x^2, \ \forall x \in [0, 1],$$

and let $(L_n)_n$ *be a sequence of positive, linear operators on* $C([0, 1])$ *such that*

$$\lim_{n \to \infty} \|L_n(p_i) - p_i\| = 0, \ i = 1, 2, 3.$$

Then for any $f \in C([0, 1])$ *the sequence* $(L_n(f))_n$ *of functions of* $C([0, 1])$ *converges uniformly to the function* f, *i.e.,*

$$\lim_{n \to \infty} \|L_n(f) - f\| = 0.$$

Proof If we consider the compact interval $K = [0, 1]$ and the functions $a_1, a_2, a_3 \in C([0, 1])$ given by

$$a_1(y) = y^2, \ a_2(y) = -2y, \ a_3(y) = 1, \ y \in [0, 1],$$

we remark that the function $(y, x) \to P_y(x) : [0, 1] \times [0, 1] \to \mathbb{R}_+$ defined by

$$P_y(x) = \sum_{i=1}^{3} a_i(y) \cdot p_i(x) = (y - x)^2$$

satisfies all requirements from Theorem 1.2.6 (Bohman–Korovkin's theorem). $\qquad \square$

Furthermore, we show the second Weierstrass approximation theorem (by trigonometric polynomials) may also be derived from Bohman–Korovkin's theorem.

First, we introduce Cesàro-Fourier operators. For any continuous 2π-periodic function f on R, we denote by $s_n(f)$ the partial Fourier sum of order n associated with function f:

$$s_n(f) = \frac{a_0}{2} + \sum_{k=1}^{n} (a_k \cdot \cos k \cdot x + b_k \cdot \sin k \cdot x),$$

where we have denoted

$$a_k = \frac{1}{\pi} \cdot \int_{-\pi}^{\pi} f(t) \cdot \cos k \cdot t dt, \quad b_k = \frac{1}{\pi} \cdot \int_{-\pi}^{\pi} f(t) \cdot \sin k \cdot t dt, \quad k \in \overline{0, n}.$$

Replacing a_k, b_k in the above definition of $S_n(f)$ and using the well-known formula:

$$\frac{1}{2} + \cos a + \cos 2 \cdot a + \ldots + \cos n \cdot a = \frac{\sin(2n+1) \cdot \frac{a}{2}}{2 \sin \frac{a}{2}}$$

we obtain

$$S_n(f) = \frac{1}{\pi} \cdot \int_{-\pi}^{\pi} f(t)$$

$$\cdot \left(\frac{1}{2} + \cos t \cdot \cos x + \ldots + \cos n \cdot t \cdot \cos n \cdot x + \sin t \cdot \sin x + \ldots + \sin n \cdot t \cdot \sin n \cdot x \right) dt$$

$$= \frac{1}{\pi} \cdot \int_{-\pi}^{\pi} f(t) \cdot \left(\frac{1}{2} + \cos(t - x) + \ldots + \cos n \cdot (t - x) \right) dt$$

$$= \frac{1}{2 \cdot \pi} \int_{-\pi}^{\pi} f(t) \cdot \frac{\sin(2 \cdot n + 1) \cdot \frac{t-x}{2}}{\sin \frac{t-x}{2}} dt.$$

Certainly S_n is a linear operator on the space $C([-\pi, \pi])$, but it is not a positive one. By contrary, the following linear operators σ_n, so-called Cèsaro-Fourier sums, given by

$$\sigma_n = \frac{s_0 + s_1 + \ldots + s_{n-1}}{n}$$

are positive.

Indeed, from the above expression of s_n and using the formula:

$$\sin \frac{a}{2} + \sin \frac{3 \cdot a}{2} + \ldots + \sin \frac{(2n-1) \cdot a}{2} = \frac{\sin^2 \frac{n \cdot a}{2}}{\sin \frac{a}{2}},$$

we get

$$\sigma_n(f)(x) = \frac{1}{2n \cdot \pi} \int_{-\pi}^{\pi} f(t) \cdot \frac{\sin^2 \frac{n \cdot (t-x)}{2}}{\sin^2 \frac{t-x}{2}} dt.$$

From the last formula, we deduce that Cesàro-Fourier sums are linear and positive operators on the space $K^* \cong \mathbb{R}/\sim$, where the equivalence relation on $\mathbb{R}$ is given by

$$x \sim y \text{ if } x - y = 2n\pi, \; n \in \mathbb{Z}.$$

This compact space K^* may be thought as the boundary of the unit disk in $\mathbb{R}^2$:

$$\left\{(x, y) \in \mathbb{R}^2; \; x^2 + y^2 = 1\right\} = \{(\cos\theta, \sin\theta); \; \theta \in [-\pi, \pi]\}$$

endowed with the trace topology or with the interval $[-\pi, \pi)$ in which the base of neighbourhoods of $-\pi$ is the family of subsets of the form:

$$\left[-\pi, -\pi + \frac{1}{n}\right) \cup \left(\pi - \frac{1}{n}, \pi\right), \; n \in \mathbb{N}^*.$$

Corollary 1.2.8 (Weierstrass [43]) *Let* $f : \mathbb{R} \to \mathbb{R}$ *be a continuous 2π-periodic function. Then, for any* $\varepsilon \in \mathbb{R}$, $\varepsilon > 0$ *there exists a trigonometric polynomial* t_ε *such that*

$$\|f - t_\varepsilon\| \leq \varepsilon.$$

We consider the compact space $K^* = [-\pi, \pi)$. The 2π-periodic functions on $\mathbb{R}$ may be identified with their restrictions to K^*. On the space $C(K^*)$ we consider the linear and positive operators σ_n, $n \in \mathbb{N}$, as before.

For the functions f_1, f_2, f_3 from K^* given by

$$f_1(x) = 1, \; f_2(x) = \cos x, \; f_3(x) = \sin x, \; x \in K^*$$

we have, by elementary calculus,

$$\sigma_n(f_1) = f_1, \; \sigma_n(f_2) = \frac{n-1}{n} \cdot f_2, \; \sigma_n(f_3) = \frac{n-1}{n} \cdot f_3, \; \forall n \in \mathbb{N}^*,$$

and therefore

$$\lim_{n \to \infty} \|\sigma_n(f_i) - f_i\| = 0, \; i = 1, 2, 3.$$

If we consider now the functions a_1, a_2, a_3 from $C(K^*)$ given by

$$a_1 = f_1, \; a_2 = -f_2, \; a_3 = -f_3,$$

then for any $x, y \in K^*$ we have

$$P_y(x) = a_1(y) \cdot f_1(x) + a_2(y) \cdot f_2(x) + a_3(y) \cdot f_3(x) = 1 - \cos(x - y) \geq 0,$$

and $P_y(x) = 0$ iff $x = y$. Using now Theorem 1.2.6, we deduce that the sequence of trigonometric polynomials $(\sigma_n(f))_n$ is uniformly convergent to f.

The power of Bohman–Korovkin' theorem is reflected in the following proof of Bernstein approximation theorem for continuous functions defined on compact subsets of $\mathbb{R}^n$.

Theorem 1.2.9 (Bernstein, see e.g. [32, p. 30]) *Let K be the compact subset $[0,1]^m$ of $\mathbb{R}^m$ and for any $f \in C(K)$ let $B_n(f) \in C(K)$ given by*

$$B_n(f)(x_1, \ldots, x_m) = \sum_{k_1=0}^{n} \cdots \sum_{k_m=0}^{n} f\left(\frac{k_1}{n}, \ldots, \frac{k_m}{n}\right) \cdot p_{nk_1}(x_1) \cdot \ldots \cdot p_{nk_m}(x_m),$$

where we have put

$$p_{nk}(t) = C_n^k \cdot t^k \cdot (1-t)^{n-k} = \frac{n!}{k!(n-k)!} \cdot t^k \cdot (1-t)^{n-k}, \ k \in \overline{1,n}.$$

Then the sequence of polynomial functions $(B_n(f))_n$ is uniformly convergent to f on K.

Proof We choose the following functions $f_{1i}, f_{2i}, f_{3i}, \ i \in \overline{1,m}$ from $C(K)$ given by

$$f_{1i}(x_1, \ldots, x_m) = 1, \ f_{2i}(x_1, \ldots, x_m) = x_i, \ f_{3i}(x_1, \ldots, x_m) = x_i^2, \ i \in \overline{1,m},$$

and we consider the functions $a_{1i}, \ a_{2i}, a_{3i}, \ i \in \overline{1,m}$ from $C(K)$ given by

$$a_{1i}(y_1, \ldots, y_m) = y_i^2, \ a_{2i}(y_1, \ldots, y_m) = -2 \cdot y_i, \ a_{3i}(y_1, \ldots, y_m) = 1, \ i \in \overline{1,m}.$$

Obviously, for any $x = (x_1, \ldots, x_m), \ y = (y_1, \ldots, y_m)$ from K we have

$$P_y(x) = \sum_{i=1}^{m} f_{1i}(x) \cdot a_{1i}(y) + \sum_{i=1}^{m} f_{2i}(x) \cdot a_{2i}(y) + \sum_{i=1}^{m} f_{3i}(x) \cdot a_{3i}(y) = \sum_{i=1}^{m}(y_i - x_i)^2,$$

and the relations $P_y(x) \geq 0$ for all $x, y \in K$ and $P_y(x) = 0$ iff $y = x$ are apparent.

On the other hand, as in the proof of Corollary 1.2.5, we have

$$\lim_{n \to \infty} \left\| B_n(f_{ji}) - f_{ji} \right\| = 0, \ \forall j \in \overline{1,3}, \ \forall i \in \overline{1,3}. \qquad \square$$

We finish the proof since the required conditions from Theorem 1.2.6 are fulfilled.

1.3   Stone-Weierstrass Theorem

The most important generalization of Weierstrass approximation theorems was obtained by Marshall H. Stone in 1937 (see e.g. [40]). Eleven years after he gave a simplified proof of his result in [40]. The generalization goes in two directions: one consists in replacing the interval [0, 1] by an arbitrary Hausdorff compact space K and the other by changing the base of approximants, algebraic or trigonometric polynomials, into an arbitrary algebra of continuous functions on K. Nowadays, this generalization bears the name "Stone-Weierstrass theorem" and it became an indispensable tool in analysis generally and a vital one in the study of continuous functions on a compact space.

The following lemma, obtained by Bruno Brosowski and Frank Deutsch (see Lemma 2 in [5]) plays an essential role in the proof of Stone-Weierstrass theorem.

Lemma 1.3.1 *Let K be a Hausdorff compact space and let A be an algebra of real continuous functions on K which contains the constant functions and separates the points of K. Then, for any two disjoint and closed subsets F_1, F_2 of K and any $\varepsilon \in \mathbb{R}$, $0 < \varepsilon < 1$ there exists a function $a \in A$ such that*

$$(i)\ 0 \leq a \leq 1$$

$$(ii)\ a(x) < \varepsilon, \forall x \in F_1$$

$$(iii)\ a(x) > \varepsilon,\ \forall x \in F_2.$$

Proof Let $x_1 \in F_1$, $x_2 \in F_2$ be arbitrarily chosen?. There exists a function $a_{x_1,x_2} \in A$ with the properties:

$$0 \leq a_{x_1,x_2} \leq 1,\, a_{x_1,x_2}(x_1) = 0,\, a_{x_1,x_2}(x_2) > 0.$$

Indeed, from the separating property of A, we may consider a function $f \in A$ such that $f(x_1) \neq f(x_2)$. Adding the constant function $-f(x_1)$, we obtain a function $g \in A$ such that $g(x_1) = 0$, $g(x_2) \neq 0$. The function g^2 belongs to A and $g^2(x_1) = 0$, $g^2(x_2) > 0$, $g^2 \geq 0$ on K.

We take

$$a_{x_1,x_2} = \frac{g^2}{\|g\|^2}.$$

The set $[a_{x_1,x_2} > 0] = \{x \in K;\ a_{x_1,x_2}(x) > 0\}$ is an open neighbourhood of x_2. Fixing $x_1 \in F_1$, we have

$$F_2 = \bigcup_{x_2 \in F_2} [a_{x_1,x_2} > 0],$$

and therefore, F_2 being compact, we may choose a finite number of points $x_{21}, \ldots, x_{2n} \in F_2$ such that

$$F_2 = \bigcup_{j=1}^{n} [a_{x_1,x_{2j}} > 0].$$

Obviously, the function $a_{x_1} \in \mathcal{A}$ given by

$$a_{x_1} = \frac{1}{n} \cdot \sum_{j=1}^{n} a_{x_1,x_{2j}}$$

vanishes at x_1, $0 \le a_{x_1} \le 1$ on K and it is strictly positive on F_2. Let us denote

$$\delta_{x_1} = \min\{a_{x_1}(t);\ t \in F_2\} > 0.$$

We observe that the set $\left[a_{x_1} < \frac{1}{3} \cdot \delta_{x_1}\right]$ is an open neighbourhood of x_1. Since F_1 is compact and

$$F_1 \subset \bigcup_{x_1 \in F_1} \left[a_{x_1} < \frac{1}{3} \cdot \delta_{x_1}\right],$$

we may choose a finite number of points, $x_1, \ldots, x_m \in F_1$ such that

$$F_1 \subset \bigcup_{i=1}^{m} \left[a_{x_i} < \frac{1}{3} \cdot \delta_{x_1}\right].$$

As $0 < \frac{\delta_{x_i}}{3} \le \frac{1}{3}$, we have $1 \le \frac{1}{\delta_{x_1}} < \frac{3}{\delta_{x_1}}$ and therefore there exists $p_i \in \mathbb{N}$, $p_i \ge 2$, such that

$$\frac{1}{\delta_{x_i}} < p_i < \frac{3}{\delta_{x_i}} \quad \text{or} \quad \frac{\delta_{x_i}}{3} < \frac{1}{p_i} < \delta_{x_i}.$$

Using now Theorem 1.1.1 for a given number $0 < \varepsilon < \frac{1}{2}$ and for each $i \in \overline{1, m}$, we choose a polynomial $p_i : [0, 1] \to [0, 1]$ such that $p_i(t) < \frac{\varepsilon}{m}$, if $t \in [0, \frac{\delta_{x_i}}{3}]$ and

$p_i(t) > 1 - \frac{\varepsilon}{m}$, if $t \in [\delta_{x_i}, 1]$. Then we have

$$p_i[a_{x_i}(x)] < \frac{\varepsilon}{m}, \ \forall x \in [a_{x_1} < \frac{\delta_{x_i}}{3}], \quad p_i[a_{x_i}(x)] > (1 - \frac{\varepsilon}{m}), \ \forall x \in F_2.$$

Then the function $a \in \mathcal{A}$ given by

$$a(x) = p_1[a_{x_1}(x)] \cdot p_2[a_{x_2}(x)] \cdot \ldots \cdot p_m[a_{x_m}(x)], \quad \forall x \in K,$$

satisfied all required conditions. Indeed, $a(x) \in [0, 1]$ for all $x \in K$, $\left(1 - \frac{\varepsilon}{m}\right)^m < a(x)$ for all $x \in F_2$, and therefore $1 - \varepsilon < \left(1 - \frac{\varepsilon}{m}\right)^m < a(x)$, $\forall x \in F_2$. As for a point $x \in F_1$, we find $i \in \overline{1, m}$ such that $x \in [a_{x_i} < \frac{\delta_{x_i}}{3}]$ and then $p_i[a_{x_i}(x)] < \frac{\varepsilon}{m} < \varepsilon$, and therefore $a(x) < \varepsilon$ for all $x \in F_1$. □

Theorem 1.3.2 (Stone-Weierstrass [5]) *Let K be a Hausdorff compact space and let $\mathcal{A}$ be an algebra of real continuous functions on K which contains the constant functions and separates the points of K. Then $\mathcal{A}$ is dense in $C(K)$ with respect to the uniform norm, i.e.,*

$$\overline{\mathcal{A}} = C(K).$$

Proof We show that for any $f \in C(K)$ and any $\varepsilon > 0$ there exists $a \in \mathcal{A}$ such that $\|f - a\| < \varepsilon$.

Since the positive f_+ and negative f_- parts of a continuous function f on K are also continuous and bounded functions on K, it is sufficient to prove the assertion only for the function $f \in C(K)$, $0 \le f \le 1$. In this case, for any $n \in \mathbb{N}^*$ we denote

$$F_i = \left[f \ge \frac{i}{2^n} \right] = \left\{ x \in K; \ f(x) \ge \frac{i}{2^n} \right\}, \ i \in \{0, 1, 2, \ldots, 2^n\},$$

and we remark that we have

$$F_0 = K \supset F_1 \supset \overset{\circ}{F_1} \supset F_2 \supset \overset{\circ}{F_2} \supset \ldots F_i \supset \overset{\circ}{F_i} \supset F_{i+1} \supset \ldots \supset F_{2^n-1} \supset \overset{\circ}{F_{12^n-1}} \supset F_{2^n}$$

and if $f(x) \in \left[\frac{i}{2^n}, \frac{i+1}{2^n} \right)$ then we have $x \in F_j$, $j \in \overline{0, i}$, $x \notin F_j$, $j \in \overline{i+1, n}$ and therefore

$$\frac{1}{2^n} \cdot \sum_{j=1}^{2^n} 1_{F_j}(x) = \frac{i}{2^n} \le f(x) < \frac{i+1}{2^n} = \frac{1}{2^n} \cdot \sum_{j=0}^{2^n} 1_{F_j}(x).$$

If we denote $u_n = \frac{1}{2^n} \cdot \sum_{j=1}^{2^n} 1_{F_j}$, $v_n = \frac{1}{2^n} \cdot \sum_{j=0}^{2^n} 1_{F_j}$, we have

$$u_n \le f \le v_n = u_n + \frac{1}{2^n}, \quad 0 \le f - u_n \le \frac{1}{2^n}, \quad 0 \le v_n - f \le \frac{1}{2^n},$$

i.e., the sequences $(u_n)_n$, $(v_n)_n$ of functions on K are uniformly convergent to f.

Since the closed sets F_i and $K \backslash \overset{\circ}{F}_{i-1}$ are disjoint, for any $\varepsilon > 0$ we may consider a function $a_i \in \mathcal{A}$ such that

$$0 \le a_i \le 1, \quad a_i(x) > 1 - \varepsilon, \; \forall x \in F_i, \quad a_i(x) < \varepsilon, \; \forall x \in K \backslash \overset{\circ}{F}_{i-1}.$$

Hence for any $\varepsilon \in (0, 1)$ there exists a function $a_i \in \mathcal{A}$ such that

$$0 \le a_i \le 1, 1_{F_i} - \varepsilon \le a_i \le 1_{F_{i-1}} + \varepsilon.$$

Further we have

$$u_n - \varepsilon = \frac{1}{2^n} \cdot \sum_{i=1}^{2^n} 1_{F_i} - \varepsilon = \frac{1}{2^n} \cdot \sum_{i=1}^{2^n} (1_{F_i} - \varepsilon) \le \frac{1}{2^n} \cdot \sum_{i=1}^{2^n} a_i \le \frac{1}{2^n} \cdot \sum_{i=1}^{2^n} (1_{F_{i-1}} + \varepsilon)$$

$$= \frac{1}{2^n} \cdot \sum_{i=1}^{2^n} 1_{F_{i-1}} + \varepsilon = v_n + \varepsilon$$

$$u_n - \varepsilon < u_n \le f \le v_n < v_n + \varepsilon = \frac{1}{2^n} + u_n + \varepsilon,$$

and therefore

$$\left\| f - \frac{1}{2^n} \cdot \sum_{i=1}^{2^n} a_i \right\| \le v_n - u_n + 2 \cdot \varepsilon < \frac{1}{2^n} + 2 \cdot \varepsilon. \qquad \square$$

The proof is finished since $\varepsilon > 0$ and $n \in \mathbb{N}$ are arbitrary, and the function $\frac{1}{2^n} \cdot \sum_{i=1}^{2^n} a_i$ belongs to $\mathcal{A}$.

Remark 1.3.3 Stone-Weierstrass theorem extends the previous Weierstrass theorems (1.1.3 and 1.1.6) as well as the Berstein theorem for compacts in $\mathbb{R}^n$ (Theorem 1.2.9).

Indeed, the set of polynomial functions on the interval $[0, 1]$ as well as the set of trigonometric polynomials on the interval $[-\pi, \pi]$ are both algebras of continuous functions containing the constant functions and separating the points of $[0, 1]$ and $[-\pi, \pi]$ for the trigonometric polynomials.

The same argument justifies the density of the polynomial functions from $\mathbb{R}^n$ in the set of $C(K)$ of all real continuous functions defined on a compact subset of $\mathbb{R}^n$.

Further we define the Uryson family of functions and also some of its applications will be presented. We remember that a topological space is called a normal space if for any non-empty closed and disjoint subsets F_1, F_2 there exist two open disjoint subsets G_1, G_2 such that $F_1 \subset G_1$, $F_2 \subset G_2$.

It is well known that any Hausdorff compact space K is a normal space as well as any metrisable space. The famous Uryson theorem gives the following characterization of normal spaces:

A topological space X is normal iff for any non-empty closed and disjoint subsets F_1, F_2 there exists a continuous function f on X with values in the interval $[0, 1]$ such that

$$f(x) = 0, \ \forall x \in F_1; \quad f(x) = 1, \ \forall x \in F_2.$$

This is the origin of Uryson family of continuous functions on a normal space introduced by I. Bucur [7] in order to unify the proofs of different density theorems.

Definition 1.3.4 A family $\mathcal{U}$ of continuous real functions on a normal space X with values in $[0, 1]$ is called Uryson family if for any closed, disjoint subsets F_1, F_2 of X and any $\varepsilon \in \left(0, \frac{1}{2}\right)$ there exists a function $u \in \mathcal{U}$ such that

$$u(x) \le \varepsilon, \forall x \in F_1; u(x) \ge 1 - \varepsilon, \forall x \in F_2.$$

Remark 1.3.5 Let K be a Hausdorff compact space and let $\mathcal{A}$ be an algebra of real continuous functions on K containing the constant functions and separating the points of K. Then, the family $\mathcal{A}_1$ given by

$$\mathcal{A}_1 = \{a \in \mathcal{A}; \ 0 \le a \le 1\}$$

is an Uryson family on K.

The assertion follows from Lemma 1.3.1.

Theorem 1.3.6 *If $\mathcal{U}$ is an Uryson family of functions on a normal space X, then the convex covering of this family, co($\mathcal{U}$), is dense in the set $C(X, [0, 1])$ of all continuous functions on X with values in the interval $[0, 1]$ if we endow $C(X, [0, 1])$ with the distance of uniform convergence on X.*

Proof For any continuous function $f : X \to [0, 1]$, and any $n \in \mathbb{N}$, $n \geq 1$, we denote

$$F_i = \left[f \geq \frac{i}{2^n} \right] := \left\{ x \in X \, \Big| \, f(x) \geq \frac{i}{2^n} \right\}, i \in \{0, 1, 2, \ldots, 2^n\}.$$

A similar argument as in the proof of Theorem 1.3.2 shows that for any $0 < \varepsilon < \frac{1}{2}$ there exists $\psi_i \in \mathcal{U}$ such that

$$\left\| f - \frac{1}{2^n} \cdot \sum_{i=1}^{2^n} \psi_i \right\| < \frac{1}{2^n} + 2\varepsilon. \qquad \square$$

The proof is finished since the function $\frac{1}{2^n} \sum_{i=1}^{2^n} \psi_i$ belongs to $\mathrm{co}\,(\mathcal{U})$.

Remark 1.3.7 Theorem 1.3.6 generalizes Stone Weierstrass theorem.

Indeed, if $\mathcal{A}$ is an algebra of continuous functions on a Hausdorff compact space K which contains the constant functions and separates the points of K, then the set $\mathcal{A}_1$,

$$\mathcal{A}_1 = \{a \in \mathcal{A}; \ 0 \leq a \leq 1\} = \mathrm{co}\,(\mathcal{A}_1),$$

is an Uryson family of functions (see Remark 1.3.5).

By Theorem 1.3.6, any continuous function $f : K \to [0, 1]$ is uniformly approximated by a sequence of $\mathcal{A}_1$. But any real continuous function $f : K \to \mathbb{R}$ is of the form:

$$f = \alpha_1 \cdot f_1 - \alpha_2 \cdot f_2,$$

with $\alpha_1, \alpha_2 \in \mathbb{R}_+$, $f_1, f_2 \in \mathcal{A}_1$.

Lemma 1.3.8 *Let K be a compact Hausdorff space and let $\mathcal{L}$ be a vector lattice of continuous real functions on K which contains the constant functions and separates the points of K. Then the set:*

$$\mathcal{L}_1 = \{a \in \mathcal{L}; 0 \leq a \leq 1\},$$

is an Uryson family on K.

Proof Let $x, y \in K$, $x \neq y$ be two points. Since $\mathcal{L}$ separates the points of K, there exists a function $\psi_{x,y} \in \mathcal{L}$ such that $\psi_{x,y}(x) < \psi_{x,y}(y)$. Adding a constant function, eventually we may suppose $\psi_{x,y}(x) < 0 < \psi_{x,y}(y)$. After multiplication with a positive constant function, the function $\psi_{x,y}$ verifies the following inequalities:

$$\psi_{x,y}(x) < 0 < \psi_{x,y}(y), \quad \psi_{x,y}(y) > 1.$$

Using the notations $f \vee g$ (resp. $f \wedge g$) for the supremum (resp. infimum) of the functions f and g we get the function $\varphi_{x,y}$:

$$\varphi_{x,y} = \left(0 \vee \psi_{x,y}\right) \wedge 1,$$

which belongs to $\mathcal{L}$, vanishing on an open neighbourhood of the point x and is equal to 1 on an open neighbourhood of the point y.

Let now F_1, F_2 be two closed and disjoint subsets of K, and for any $x \in F_1$, $y \in F_2$, let $\varphi_{x,y} \in \mathcal{L}$ be such that $\varphi_{x,y}$ vanishes on an open neighbourhood $U^y(x)$ of x and it is constant equal 1 on an open neighbourhood $U^x(y)$ of y. If we fix $x \in F_1$, we have $F_2 \subset \bigcup_{y \in F_2} U^x(y)$, and therefore $F_2 \subset \bigcup_{i=1}^n U^x(y_i)$ for some choice of the points $y_1, \dots, y_n \in F_2$.

If we put $\varphi_x = \overset{n}{\underset{i=1}{\vee}} \varphi_{x,y_i}$, the function φ_x belongs to $\mathcal{L}$, vanishes on an open neighborhood $U(x)$ of the point x and it is equal to 1 on F_2. With the above notations, we have $F_1 \subset \bigcup_{x \in F_1} U(x)$, and using the compactness of F_1 we have $F_1 \subset \bigcup_{i=1}^m U(x_i)$ for some choice of the points $x_1, \dots, x_m \in F_1$. It is now evident that the function $\varphi = \overset{m}{\underset{i=1}{\wedge}} \varphi_{x_i}$ belongs to $\mathcal{L}, \varphi = 0$ on F_1 and $\varphi = 1$ on F_2.   □

Theorem 1.3.9 (Stone-Weierstrass [7]) *Let K be a compact Hausdorff space and let $\mathcal{L}$ be a vector lattice of continuous real functions on K which contains the constant functions and separates the points of K. Then $\mathcal{L}$ is uniformly dense in the space $C(K)$ of all real continuous functions on K.*

Indeed, for any $f \in C(K)$ we have $f = f^+ - f^-$ where $f^+ = f \vee 0$ and $f^- = (-f) \vee 0$. Obviously the functions $\frac{f^+}{\|f\|}$, $\frac{f^-}{\|f\|}$ belong to $C(K, [0, 1])$. On the other hand, since $\mathcal{L}_1 = \{a \in \mathcal{L}; 0 \le a \le 1\}$ is an Uryson family on K, just a convex one, we deduce, using Theorem 1.3.6, that the functions $\frac{f^+}{\|f\|}$, $\frac{f^-}{\|f\|}$ belong to the closure of $\mathcal{L}_1$ in $C(K, [0, 1])$. We have

$$f = \|f\| \cdot \left(\frac{f^+}{\|f\|} - \frac{f^-}{\|f\|}\right) \in \cdot \|f\| \cdot \left(\overline{\mathcal{L}_1} - \overline{\mathcal{L}_1}\right) \subset \overline{\mathcal{L}}, \quad C(K) \subset \overline{\mathcal{L}}.$$

In 1959 Louis de Branges [4] give a new proof of the Stone-Weierstrass theorem concerning the algebras of continuous functions on Hausdorff, compact spaces, using two fundamental tools in functional analysis: Hahn-Banach and Krein-Milman theorems.

Furthermore, we present this newer proof containing some ideas which may be useful in obtaining different Stone-Weierstrass type generalizations.

In this section, K will be a Hausdorff compact space, $C(K)$ the Banach space of all real continuous functions on K endowed with the uniform norm and $M(K)$ will be the dual of $(C(K), \ \|\|)$, i.e., the set of all real linear continuous functionals μ on $C(K)$.

If for any two functions $f, g \in C(K)$ we denote by

$$[f, g] = \{\varphi \in C(K); \ f(x) \leq \varphi(x) \leq g(x), \ x \in K\},$$

then the closed unit ball in the Banach space $(C(K), \ \|\|)$ is the set $[-\mathbf{1}, \mathbf{1}]$ where $\mathbf{1}$ is the constant function on K equal 1 at any point of K.

$M(K)$ endowed with the pointwise addition and multiplication with scalars from $\mathbb{R}$ is a linear space and just a Banach space if we endow $M(K)$ with the dual norm:

$$\|\mu\| = \sup\{\mu(f); \ f \in [-\mathbf{1}, \mathbf{1}]\}.$$

In fact, $M(K)$ is a Banach lattice with respect to the order relation "$\leq$" given by

$$\mu \leq \nu \ \Leftrightarrow \ \mu(f) \leq \nu(f), \ \forall f \in C^+(K),$$

where $C^+(K)$ is the convex cone of all positive functions of $C(K)$.

If we denote by $M^+(K)$ the subset of $M(K)$ given by

$$M^+(K) = \{\mu \in M(K); \ \mu \geq 0\} = \{\mu \in M(K); \ \mu(f) \geq 0, \forall f \in C^+(K)\},$$

then for any $\mu \in M(K)$ there is the smallest element in $M^+(K)$ denoted by μ^+, such that

$$\mu^+(f) = \sup\{\mu(g); \ g \in C^+(K), \ g \leq f\},$$

for any $f \in C^+(K)$. Also there exists a smallest element in $M^+(K)$ denoted by μ^- such that

$$\mu^-(f) = \sup\{-\mu(g); \ g \in C^+(K), \ g \leq f\},$$

for any $f \in C^+(K)$. Moreover we have $\mu = \mu^+ - \mu^-$ and the measure $\mu^+ + \mu^-$ is the modulus of μ denoted by $|\mu|$, i.e., the smallest element ν in $M^+(K)$ such that $|\mu(f)| \leq \nu(|f|)$, for all $f \in C(K)$. In fact we have

$$|\mu|(f) = \mu^+(f) + \mu^-(f) = \sup \ \{\mu(g); \ g \in C(K), \ |g| \leq f\},$$

for any $f \in C^+(K)$. We remember also that for any $\mu \in M^+(K)$ we have $\|\mu\| = \mu(\mathbf{1})$ and for any $\mu \in M(K)$ we have

$$\|\mu\| = \| \ |\mu| \ \| = |\mu|(\mathbf{1}).$$

Often, the elements of $M(K)$ are called Radon measures. The reason for this denomination is the fact that for any positive element μ of $M(K)$ there exists a positive measure λ_μ on $\mathscr{B}(K)$ the σ-algebra of Borel sets of K such that

$$\mu(f) = \int f d\lambda_\mu, \ \forall f \in C(K),$$

and λ_μ is uniquely determined by its regularity, i.e.,

$$\lambda_\mu(A) = \sup \left\{ \lambda_\mu(K); \ K - \text{compact}, \ K \subset A \right\},$$

for any $A \in \mathscr{B}(K)$.

For an arbitrary $\mu \in M(K)$ there exists a sign measure λ_μ on $\mathscr{B}(K)$ such that

$$\mu(f) = \int f d\lambda_\mu, \ \forall f \in C(K),$$

namely, $\lambda_\mu = \lambda_{\mu+} - \lambda_{\mu-}$. In fact $\lambda_{\mu+}$ (resp. $\lambda_{\mu-}$) is the positive (resp. negative) part of the measure λ_μ, i.e., there exists $A \subset \mathscr{B}(K)$ such that

$$\lambda_{\mu+}(A) = \lambda_{\mu+}(K), \ \ \lambda_{\mu-}(K \backslash A) = \lambda_{\mu-}(K), \ \text{and} \ \lambda_{|\mu|} = \lambda_{\mu+} + \lambda_{\mu-}.$$

The support of an element $\mu \in M(K)$ is by definition the support of the associated measure λ_μ, i.e., the smallest closed subset F of K such that $\lambda_{|\mu|}(F) = \lambda_{|\mu|}(K)$.

We denote the support of μ by $\text{Supp}(\mu)$. We remark that for any $g \in L^1(\lambda_\mu) = L^1\left(|\lambda_\mu|\right)$ the map, denoted by g_μ, defined on $C(K)$ by

$$g_\mu(f) = \int f \cdot g d\mu, \ \forall f \in C(K)$$

is linear and we have

$$\left| g_\mu(f) \right| \leq \|f\| \cdot \int |g| \, d \, |\lambda_\mu| = \|f\| \cdot \int |g| \, d \, |\lambda_{|\mu|}|, \ \forall f \in C(K),$$

i.e., this map belongs to $M(K)$. One can show that $\|g_\mu\| = \int |g| \, d\lambda_{|\mu|}$.

Further from now on we shall freely use the notation $\mu(h)$ instead of $\int h d\lambda_\mu$ whenever the former expression makes sense. Also, for any subset E of $C(K)$ we denote by E^0 the polar of E with respect to the duality $\langle C(K), \ M(K) \rangle$ namely:

$$E^0 = \{\mu \in M(K); \ \mu(f) \leq 1, \ f \in E\}.$$

If $\mathcal{C}$ is a convex cone in $C(K)$ (resp. F a linear subspace of $C(K)$) we have

$$\mathcal{C}^0 = \{\mu \in M(K); \ \mu(f) \le 0, \ \forall f \in \mathcal{C}\} \quad (\text{resp. } \mathrm{F}^0 = \{\mu \in M(K); \ \mu(f) = 0, \ \forall f \in F\}).$$

Lemma 1.3.10 (Louis de Branges [4]) *Let F be a linear subspace of $C(K)$ and let μ be an extreme point of the convex subset $F^0 \cap [-1, 1]^0$ of $M(K)$. If $g \in C(K)$ is such that $g\mu \in F^0$, then g is constant on $\mathrm{Supp}(\mu)$.*

Proof Since $0 \in [-1, 1]^0$ and $\|\mu\| \le 1$, we have necessarily $\|\mu\| = 1$, otherwise μ is an interior point of the unit ball of F^0, and, therefore, it is not an extreme point of this ball. Now, adding a positive constant function to g and multiplying this sum by a strictly positive number we can obtain a function h such that $h_\mu \in F^0$, $0 \le h \le 1$. We show that h is constant on $\mathrm{Supp}(\mu)$.

We consider the following measures μ_1, μ_2:

$$\mu_1 = \frac{\mu + h\mu}{\|\mu + h\mu\|}, \quad \mu_2 = \frac{\mu - h\mu}{\|\mu - h\mu\|}.$$

Obviously, μ_1, μ_2 belong to $F^0 \cap [-1, 1]^0$, and we have

$$\|\mu + h\mu\| = |\mu| \, (|1 + h|) = |\mu| \, (1 + h)$$

$$\|\mu - h\mu\| = |\mu| \, (|1 - h|) = |\mu| \, (1 - h)$$

$$\|\mu + h\mu\| + \|\mu - h\mu\| = |\mu| \, (1 + h + 1 - h) = |\mu| \, (2) = 2,$$

$$\frac{\|\mu + h\mu\|}{2} + \frac{\|\mu - h\mu\|}{2} = 1,$$

$$\frac{\|\mu + h\mu\|}{2} \cdot \mu_1 + \frac{\|\mu - h\mu\|}{2} \cdot \mu_2 = \mu.$$

Since μ is extremal, we have $\mu_1 = \mu_2 = \mu$. Hence

$$\mu = \frac{\mu + h\mu}{\|\mu + h\mu\|} = \frac{(1 + h)\mu}{1 + |\mu| \, (h)},$$

$$(1 + |\mu| \, (h))\mu = (1 + h)\mu, \quad h \cdot \mu = |\mu| \, (h) \cdot \mu, \quad h = |\mu| \, (h) \text{ on } \mathrm{Supp}(\mu). \qquad \square$$

Theorem 1.3.11 (Stone-Weierstrass [7]) *Let K be a Hausdorff compact space and let $\mathcal{A}$ be an algebra of real continuous functions on K which contains the constant functions* *and separates the points of K. Then, $\mathcal{A}$ is dense in $C(K)$ with respect to the uniform norm, i.e.,*

$$\overline{\mathcal{A}} = C(K).$$

Proof If $\overline{\mathcal{A}} \neq C(K)$, then there exists $g \in C(K) \backslash \overline{\mathcal{A}}$, and from Hahn-Banach theorem there exists $\mu \in M(K)$, $\|\mu\| = 1$, such that $\mu(a) = 0$, $\forall a \in \mathcal{A}$ and $\mu(g) \neq 0$. Clearly, $\mu \in \mathcal{A}^0 \cap [-1, 1]^0$. Since the subset $\mathcal{A}^0 \cap [-1, 1]^0$ is compact and convex, from Krein-Milman theorem, it results that $\mathcal{A}^0 \cap [-1, 1]^0 = \overline{co} \, Ext \left(\mathcal{A}^0 \cap [-1, 1]^0 \right)$, hence there exists $\nu \in Ext \left(\mathcal{A}^0 \cap [-1, 1]^0 \right)$, $\nu \neq 0$ such that $\nu(g) \neq 0$. We remark that for any $a \in \mathcal{A}$ we have $a\nu \in \mathcal{A}^0$ and using de Branges Lemma we deduce that a is constant on Suppν. But $\mathcal{A}$ separates the points of K and therefore the support of ν is a singleton, i.e., Supp$(\nu) = \{x_0\}$ and so $\nu = \nu(1) \cdot \varepsilon_{x_0}$. Since $\nu \in \mathcal{A}^0$ and $1 \in \mathcal{A}$, we get $\nu(1) = 0$, and so $\nu = 0$ which contradicts the relation $\nu \neq 0$. This contradiction comes from the initial hypothesis $\overline{\mathcal{A}} \neq C(K)$. $\square$

Remark 1.3.12 The conclusion of Theorem 1.3.11 fails in the case of functions with complex values.

Indeed, let $D = \{z \in \mathbb{C}; |z| \leq 1\}$ and let H be the algebra of all continuous complex valued functions on D which are holomorphic in the interior of D. Obviously, H contains the constant functions and separates the points of D, since it contains all complex polynomials. Moreover it is closed with respect to the uniform convergence topology, but $H \neq C(D, \mathbb{C})$, since the real continuous functions on D do not belong to H.

Nevertheless, the conclusion in the Stone–Weierstrass theorem is still valid in the complex case if the algebra $\mathcal{A}$ is self-adjoint, i.e., for any element $a \in \mathcal{A}$, its complex conjugate $\bar{a}$ belongs to $\mathcal{A}$.

Remark 1.3.13 If $\mathcal{A}$ is a self-adjoint algebra, and we denote by $Re(\mathcal{A}) = \{Re(a); \; a \in \mathcal{A}\}$, then we have

(i) $Re(\mathcal{A}) \subset \mathcal{A}$, (ii) $Re(\mathcal{A})$ is an algebra, (iii) $\mathcal{A} = Re(\mathcal{A}) + i \cdot Re(\mathcal{A})$.

Indeed, for any $a \in \mathcal{A}$ we have $Re(a) = \frac{a + \bar{a}}{2} \in \mathcal{A}$ and therefore $Re(\mathcal{A}) \subset \mathcal{A}$. Clearly $Re(\mathcal{A})$ is a linear real vector space and $Re(\mathcal{A})$ is an algebra, because if $u_1, u_2 \in Re(\mathcal{A})$ then $u_1 \cdot u_2 \in \mathcal{A}$ and $u_1 \cdot u_2 = Re(u_1 \cdot u_2 + i \cdot 0) \in Re(\mathcal{A})$. Since for any $f \in \overline{\mathcal{A}}$, we have $Im(f) = Re(-i \cdot f) \in \mathcal{A}$ we get that $Im(f) \in Re(\mathcal{A})$, and therefore $\mathcal{A} = Re(\mathcal{A}) + i \cdot Re(\mathcal{A})$.

With these remarks we are able to establish the Stone-Weierstrass theorem for complex functions.

Theorem 1.3.14 (Stone-Weierstrass [10]) *Let K be a Hausdorff compact space and let $\mathcal{A}$ be an algebra (over the field of complex numbers) of continuous complex valued functions on K which is self-adjoint, contains the constant functions and separates the* *points of K. Then A is dense in C(K, C)-the set of all continuous complex valued functions on K, with respect to the uniform norm, i.e.,*

$$\overline{\mathcal{A}} = C(K, \mathrm{C}).$$

Proof By the above remarks, and by hypotheses, the real algebra $Re(\mathcal{A})$ contains the constant functions and separates the points of K. Hence $Re(\mathcal{A})$ is dense in $C(K) = C(K, \mathrm{R})$. Further we have

$$C(K, \mathrm{C}) = C(K) + i \cdot C(K)$$

$$= \overline{Re(\mathcal{A})} + i \cdot \overline{Re(\mathcal{A})} \subset \overline{Re(\mathcal{A}) + i \cdot Re(\mathcal{A})} = \overline{\mathcal{A}}. \qquad \square$$

1.4   Bishop Type Theorems

Erett Bishop (see [2]) generalizes the Stone-Weierstrass theorem for the case of non-selfadjoint algebras of continuous complex valued functions. To present his results we need some preliminaries.

Definition 1.4.1 Let K be a Hausdorff compact space and let $\mathcal{A}$ be an algebra of continuous complex valued functions on K. A subset S of K is called antisymmetric with respect to $\mathcal{A}$ or $\mathcal{A}$-antisymmetric, if any element $a \in \mathcal{A}$ is constant on S if $Im\, a \equiv 0$ on S.

Remark 1.4.2 A subset $S \subset K$ is $\mathcal{A}$-antisymmetric iff any element $a \in \mathcal{A}$ is constant on S if $\bar{a}\,|S \in \mathcal{A}\,|S$, i.e., there exists $a' \in \mathcal{A}$ such that $\bar{a}\,|S = a'\,|S$ or equivalently $a\,|S = \bar{a}'\,|S$, where for any complex number z we denote by $\bar{z}$ its conjugate.

Indeed, if $a\,|S = \overline{a'}\,|S$ with $a' \in \mathcal{A}$, then $Im(a + a') = 0$ on S, and since S is $\mathcal{A}$-antisymmetric, it follows that the function $a + a'$ is constant on S.

But $Re(a) = Re(a') = \frac{1}{2} \cdot Re(a + a')$ on S, hence $Re(a)$ is constant on S. We have $i \cdot a\,|S = i \cdot a'\,|S = \overline{(-i \cdot a')}\,|S \in \mathcal{A}\,|S$, and therefore the function $Re(i \cdot a) = -Im(a)$ is constant on S.

If any element $a \in \mathcal{A}$ is constant on S if $\bar{a}\,|S \in \mathcal{A}\,|S$, then for any $a \in \mathcal{A}$ which is real on S we have $\bar{a}\,|S = a\,|S$ and therefore a is constant on S.

Example 1.4.3 Let K be a compact subset of the complex plane C and let $\mathcal{H}$ be the algebra of all continuous complex functions on K which are holomorphic in $\overset{\circ}{K}$-the interior of K. Then any component E of $\overset{\circ}{K}$ is $\mathcal{H}$-antisymmetric.

We remember that E is called component of a subset M of C if E is connected, $E \subset M$ and there is no other connected subset of M larger than E. If M is an open subset of C, then any component E of M is open. Indeed, let $x_0 \in E$ and let $B(x_0, r)$, $r > 0$ a ball included in the open set M. The sets $B(x_0, r)$ and E being connected and $E \cap B(x_0, r) \neq \phi$ it follows that $E \cup B(x_0, r)$ is connected. Since we have

$$E \subset E \cup B(x_0, r) \subset M,$$

we deduce $E = E \cup B(x_0, r)$ and hence $x_0 \in \overset{\circ}{E}$.

In the above case, if $E \subset \overset{\circ}{K}$ is a component of $\overset{\circ}{K}$ and if $h \in \mathcal{H}$ is such that $h \, | E$ is real, we deduce that $Im \, h = 0$, and therefore the complex derivative of h on the connex, open set E is 0, i.e., h is constant on E.

Coming back to the general case where K is an arbitrary Hausdorff compact space and $\mathcal{A} \subset C(K, \mathbb{C})$ is an arbitrary algebra, we shall denote by Σ the set of all antisymmetric with respect to $\mathcal{A}$ subsets of K. Just Definition 1.4.1 it follows:

Remark 1.4.4 The family Σ has the following properties:

 (i) $\{x_0\} \in \Sigma$, $\forall x_0 \in K$.
 (ii) If $S_1, S_2 \in \Sigma$, and $S_1 \cap S_2 \neq \phi$, then $S_1 \cup S_2 \in \Sigma \in \sum$.
 (iii) If $S \in \Sigma$, then its closure $\overline{S} \in \Sigma$.
 (iv) Any point $x \in K$ belongs to a maximal $\mathcal{A}$-antisymmetric subset denoted by S_x.
 (v) $K = \cup \{S_x; \ x \in K\}$, where for any $x, y \in K$, $x \neq y$, we have either $S_1 \cap S_2 = \phi$, or $S_x = S_y$.

Further we shall use the following notations:

For any closed subset F of K and any $f \in C(X, \mathbb{C})$, we denote $f \, | F$ the restriction of f to F and $\|f\|_F = \sup \{ \ |f(x)| \ ; \ x \in F\} = \|f \, | F \|$. Also, for any non-empty subset M of $C(X, \mathbb{C})$ we denote by $dist(f \, | F, \ M \, | F)$ the distance of $f \, | F$ up to $M \, | F$, i.e.,

$$dist \, (f \, | F, \ M \, | F) = \inf \{ \|f - m\|_F; \ m \in M \} .$$

For $F = K$ we have

$$dist(f, M) = \inf \{ \|f - m\|; \ m \in M \} .$$

Lemma 1.4.5 *The family of closed subsets F of K with the property:*

$$dist(f \, | F, \ M \, | F) = dist \, (f, M) ,$$

has at least a minimal element with respect to the inclusion relation.

Proof Let $\mathcal{F}$ be the family of all closed subsets F of K such that

$$dist(f\,|F\,,\,M\,|F\,) = dist\,(f, M)\,,$$

and let $\mathcal{F}_0$ be a subfamily of $\mathcal{F}$ totally ordered by inclusion. For any $m \in M$ and any $F \in \mathcal{F}_0$ we denote

$$F^m = \{x \in F;\ |f(x) - m(x)| \geq d\}\,,\quad d = dist\,(f, M)\,.$$

Since for any $F_1, F_2 \in \mathcal{F}_0$ we have $F_1 \subset F_2$ or $F_2 \subset F_1$, we get

$$F_1{}^m \subset F_2{}^m\ \text{or}\ F_2{}^m \subset F_1{}^m,$$

and therefore the family $(F^m)_{F \in \mathcal{F}_0}$ is totally ordered by inclusion. Just from the definition of F^m and using Weierstrass theorem relative to the upper bound of a continuous function on a compact space, we deduce that $F^m \neq \phi$ for all $F \in \mathcal{F}_0$.

Since F^m is compact for any $F \in \mathcal{F}_0$ we get $\bigcap \{F^m;\ F \in \mathcal{F}_0\} \neq \phi$, and therefore

$$\left\{ x \in \bigcap_{F \in \mathcal{F}_0} F;\ |f(x) - m(x)| \geq d \right\} = \bigcap_{F \in \mathcal{F}_0} F^m \neq \phi,\ m \in M.$$

Denoting $F_0 = \cap \{F;\ F \in \mathcal{F}_0\}$, we get

$$dist\,(f\,|F_0,\ M\,|F_0\,) \geq d\ i.e.\ dist\,(f\,|F_0,\ M\,|F_0\,) = dist\,(f, M)\,,\ F_0 \in \mathcal{F}. \qquad \square$$

Theorem 1.4.6 (Machado [18]) *Let $\mathcal{A} \subset C(K, \mathbb{C})$ be an algebra and let f be a continuous complex valued function on K. Then, there exists an antisymmetric S with respect to $\mathcal{A}$ such that*

$$dist\,(f, \mathcal{A}) = dist\,(f\,|S,\ \mathcal{A}\,|S\,)\,.$$

Proof From Lemma 1.4.5, we may choose a minimal element S with respect to the inclusion order relation in the set $\mathcal{F}$ of all closed subsets F of K such that

$$\|f - \mathcal{A}\|_F = dist\,(f\,|F, \mathcal{A}\,|F\,) = dist\,(f, \mathcal{A})\,.$$

We show that S is an antisymmetric subset with respect to $\mathcal{A}$. For this, we consider an element $a \in \mathcal{A}$ such that the restriction of a to S is a real function. We want to show that a is constant on S.

On the contrary case, we consider x_0, $y_0 \in S$ such that $a(x_0) \neq a(y_0)$. We may suppose that $a(x_0) = 0$ or $a(y_0) = 0$. Indeed, if $a(x_0) \neq 0$ and $a(y_0) \neq 0$, then the function $a' \in \mathcal{A}$:

$$a'(x) = [a(x_0) - a(x)] \cdot a(x), \ \forall x \in K,$$

satisfies the desired conditions because $a'(x_0) = 0$ and $a'(y_0) = [a(x_0) - a(y_0)] \cdot a(y_0) \neq 0$.

We observe that the function $a_0 \in \mathcal{A}$ given by $a_0 = \frac{a'^2}{\|a'^2\|_s}$ is positive, $a_0(x_0) = 0$, $0 \leq a_0 \leq 1$, and $\|a_0\| = 1$. Furthermore, we denote

$$Y = \left[a_0 \leq \frac{2}{3} \right] - \left\{ x \in S; a_0(x) \leq \frac{2}{3} \right\}, \ Z = \left[a_0 \geq \frac{1}{3} \right] = \left\{ x \in S; a_0(x) \geq \frac{1}{3} \right\}.$$

Obviously, we have

$$S = Y \cup Z = (Y \backslash Z) \cup (Y \cap Z) \cup (Z \backslash Y), Y \neq S, \ Z \neq S,$$

$$a_0(x) < \frac{1}{3}, \ \forall x \in Y \backslash Z; \ \ \frac{1}{3} \leq a_0(x) \leq \frac{2}{3}, \ \forall x \in Y \cap Z; \ \ a_0(x) > \frac{2}{3}, \ \forall x \in Z \backslash Y.$$

Since S is an element minimal in Σ, we get $Y \notin \Sigma$, $Z \notin \Sigma$. Hence there are two functions $a_Y, a_Z \in \mathcal{A}$ such that

$$\|f - a_Y\| < d, \ \ \|f - a_Z\| < d.$$

From Theorem 1.1.1, the sequence $\varphi_n : [0, 1] \to [0, 1]$ of functions given by

$$\varphi_n(t) = \left(1 - t^n \right)^{3^n}$$

is uniformly convergent to 1 on the interval $\left[0, \frac{1}{3} \right]$ and is uniformly convergent to 0 on the interval $\left[\frac{2}{3}, 1 \right]$, and consequently the sequence $(a_n)_n$ given by $a_n(x) = \varphi_n[a_0(x)]$ is uniformly convergent to 1 on the set $Y \backslash Z$ and is uniformly convergent to 0 on the set $Z \backslash Y$.

If we denote $b_n = 1 - a_n$, $n \geq 1$, then $(b_n)_n$ converges uniformly to 0 on $Y \backslash Z$ (or to 1 on $Z \backslash Y$). Hence the sequence $(a_Y \cdot a_n + a_Z \cdot b_n)_n$ converges uniformly to a_Y on $Y \backslash Z$ (or to a_Z on $Z \backslash Y$) and therefore for n sufficiently large we have

$$|f(x) - [a_n(x) \cdot a_Y(x) + b_n(x) \cdot a_Z(x)]| < d, \ \forall x \in (Y \backslash Z) \cup (Z \backslash Y).$$

If $x \in Y \cap Z$, then we have

$$|f(x) - a_Y(x)| \leq \|f - a_Y\| < d, \ \text{and} \ |f(x) - a_Z(x)| \leq \|f - a_Z\| < d.$$

Hence if $x \in Y \cap Z$, we have

$$|f(x) - [a_n(x) \cdot a_Y(x) + b_n(x) \cdot a_Z(x)]|$$

$$= |a_n(x) \cdot [f(x) - a_Y(x)] + b_n(x) \cdot [f(x) - a_Z(x)]|a_n(x)$$

$$\cdot |f(x) - a_Y(x)| + b_n(x) \cdot |f(x) - a_Z(x)| < a_n(x) \cdot d + b_n(x) \cdot d = d.$$

From the preceding considerations, we have

$$\left\| f - [a_n \cdot a_y + b_n \cdot a_Z] \right\|_S < d, \quad a_n \cdot a_Y + b_n \cdot a_Z \in \mathcal{A},$$

which contradicts the fact that $S \in \mathcal{F}$. Hence S is $\mathcal{A}$-antisymmetric. $\qquad\square$

Theorem 1.4.7 (Bishop [2]) *Let $\mathcal{A} \subset C(K, \mathbb{C})$ be a subalgebra and let f be an element of $C(K, \mathbb{C})$. Then we have $f \in \overline{\mathcal{A}}$, iff $f \,|S \in \overline{\mathcal{A}\,|S}$ for any maximal antisymmetric set S with respect to $\mathcal{A}$.*

Proof Using Theorem 1.4.6, there exists an antisymmetric set with respect to $\mathcal{A}$, S_0 such that

$$dist\,(f, \mathcal{A}) = dist\,(f\,|S_0, \mathcal{A}\,|S_0)\,.$$

But by hypothesis $dist\,(f\,|S_0, \ \mathcal{A}\,|S_0) = 0$, and therefore $dist\,(f, \ \mathcal{A}) = 0$, hence $f \in \overline{\mathcal{A}}$. $\qquad\square$

Remark 1.4.8 If $\mathcal{A}$ is a self-adjoint subalgebra of $C(K, \mathbb{C})$ then for any antisymmetric set S with respect to $\mathcal{A}$ the functions $a \in \mathcal{A}$ are constant on S (we say that S is a constant set for $\mathcal{A}$).

The assertion follows directly from Remark 1.4.4. In fact, in this case, a set S is antisymmetric with respect to $\mathcal{A}$ iff it is a constant set for $\mathcal{A}$.

Giving an algebra $\mathcal{A} \subset C(K, \mathbb{C})$, for any $x \in K$ we denote

$$[x] = \{y \in K;\, a(x) = a(y),\, \forall a \in \mathcal{A}\}\,.$$

Obviously, $[x]$ is the greatest constant set for $\mathcal{A}$ which contains x, and if $x, y \in K$ we have either $[x] = [y]$ or $[x] \cap [y] = \phi$. The family $\{[x];\ x \in K\}$ of closed subsets of K is a partition of K, i.e., $K = \cup\{[x];\ x \in K\}$.

Corollary 1.4.9 *If $\mathcal{A}$ is a self-adjoint subalgebra of $C(K, \mathbb{C})$, then the following assertions are equivalent:*

$$(i) \quad f \in \overline{\mathcal{A}},$$

$$(ii) \quad f \mid [x] \in \overline{\mathcal{A} \mid [x]}, \; \forall x \in K,$$

$$(iii) f \mid [x] \in \mathcal{A} \mid [x], \; \forall x \in K$$

Indeed, the equivalence $(i) \Leftrightarrow (ii)$ follows from Theorem 1.4.7 and the coincidence between the antisymmetric and the constant sets with respect to $\mathcal{A}$. The equivalence $(ii) \Leftrightarrow (iii)$ may be deduced from the equality $\mathcal{A} \mid [x] = \overline{\mathcal{A} \mid [x]}$.

Corollary 1.4.10 (Stone-Weierstrass [10]) *If $\mathcal{A}$ is a self-adjoint subalgebra of $C(K, \mathbb{C})$ which contains the constant functions and separates the points of K, then $\mathcal{A}$ is uniformly dense in $C(K, \mathbb{C})$, i.e.,*

$$\overline{\mathcal{A}} = C(K, \mathbb{C}).$$

Proof Since $\mathcal{A}$ separates the points of K, for any $x \in K$ the set $[x]$ is a singleton, i.e. $[x] = \{x\}$. On the other hand, since the constant functions belong to $\mathcal{A}$, we have

$$\mathcal{A} \mid [x] = \{\varphi_c; \; c \in \mathbb{C}\}, \; \varphi_c : \{x\} \to \mathbb{C}, \; \varphi_c(x) = c.$$

The proof is now finished if we apply the previous corollary.   □

As an application of Bishop's theorem, we present the following Rudin's result.

Theorem 1.4.11 *Let K be a compact subset of $I^n \times \mathbb{C}$ where $I = [0, 1]$, such that for any $t \in I^n$ the section $K_t = \{z \in \mathbb{C}; \; (t, z) \in K\}$ does not separate $\mathbb{C}$, that is the set $\mathbb{C} \backslash K_t$ is connected.*

For any $g \in C(K, \mathbb{C})$, we define the function $g_t : K_t \to \mathbb{C}$ by $g_t(z) = g(t, z)$. We suppose now that the previous function has the property that for any $t \in I^n$ the associated function g_t is holomorphic on $\overset{\circ}{K_t}$-the interior of K_t. Then for any $\varepsilon > 0$ there exists a polynomial function P such that

$$|g(t, z) - P(t, z)| < \varepsilon, \quad \forall (t, z) \in K.$$

Proof We denote by $\mathcal{A}$ the algebra of all polynomials defined on K with values in $\mathbb{C}$. Since for any $t_1, t_2 \in I^n$ the polynomial $(t, z) \to \|t - t_1\|^2$ separates the points t_1, t_2, we deduce that for any $\mathcal{A}$-antisymmetric set S of K there exists $t_0 \in I^n$ such that $S \subset \{t_0\} \times K_{t_0}$.

To show that $g \in \overline{\mathcal{A}}$, it will be sufficient to prove that for any $\mathcal{A}$-antisymmetric set S of K we have $g \mid S \in \overline{\mathcal{A} \mid S}$. For such set S let the point $t_0 \in I^n$ such that $S \subset \{t_0\} \times K_{t_0}$. By hypothesis, the function $g_{t_0} : K_{t_0} \to \mathbb{C}$, given by $g_{t_0}(z) = g(t_0, z)$, is continuous on K_{t_0} and holomorphic on the interior of K_{t_0}. So, by Mergelyan theorem there exists a sequence of polynomials $(P_n)_n$ on $\mathbb{C}$ such that $(P_n)_n$ is uniformly convergent to g_{t_0} on K_{t_0}. We observe now that the sequence $(Q_n)_n$ of polynomials on K given by $Q_n(t, z) = P_n(z)$ converges uniformly to the function g on the set S of K, i.e., $g \mid S \in \overline{\mathcal{A} \mid S}$. □

1.5   Sets of Functions with VN Property (von Neumann Property)

Definition 1.5.1 We say that a family M of real functions defined on the arbitrary set X, containing the constant functions 0 and 1 is φ-convex, where $\varphi \in M$, if for any $f, g \in M$ we have

$$\varphi \cdot f + (1 - \varphi) \cdot g \in M.$$

We remark that $1 - \varphi \in M$ since $1 - \varphi = \varphi \cdot 0 + (1 - \varphi) \cdot 1$ and so M is also $(1 - \varphi)$-convex. Moreover M is stable by multiplication with φ since $\varphi \cdot f = \varphi \cdot f + (1 - \varphi) \cdot 0$.

Lemma 1.5.2 *If M is φ-convex, then $\varphi^n \in M$ for any $n \in \mathbb{N}^*$ and M is φ^n-convex. Hence M is $(1 - \varphi^n)$-convex and also $(1 - \varphi^n)^m$-convex for any $n, m \in \mathbb{N}^*$.*

Proof Since $\varphi \cdot f \in M$ for all $f \in M$ we get $\varphi^2 \cdot f, \varphi^3 \cdot f, \ldots, \varphi^n \cdot f \in M$. Particularly, taking $f = 1$ we have $\varphi^n \in M$. We show now inductively that M is φ^n-convex for any $n \in \mathbb{N}^*$.

We suppose that M is φ^k-convex, i.e.,

$$\varphi^k \cdot f + (1 - \varphi^k) \cdot g \in M, \quad \forall f, g \in M.$$

From hypothesis, we have

$$\varphi^{k+1} \cdot f + (1 - \varphi^{k+1}) \cdot g = \varphi \cdot [\varphi^k \cdot f + (1 - \varphi^k) \cdot g] + (1 - \varphi) \cdot g \in M,$$

for any $f, g \in M$, i.e., M is φ^{k+1}-convex. □

Definition 1.5.3 We say that a subset $M \subset C(X, [0, 1])$ possesses the VN property if we have

$$\varphi \cdot f + (1 - \varphi) \cdot g \in M, \forall \varphi, f, g \in M.$$

We observe that if the constant functions 0 and 1 belong to M, then M possesses the VN property iff M is φ-convex for any $\varphi \in M$.

Remark 1.5.4 If M possesses the VN property and $0 \in M$ then $\varphi \cdot f \in M$, $\forall \varphi, f \in M$ and $\varphi^n \in M$, $\forall \varphi \in M$, $\forall n \in \mathbb{N}^*$. If the constant functions 0 *and* $1 \in M$, then $1 - \varphi \in M$, $\forall \varphi \in M$ and $(1 - \varphi^n)^m \in M$, $\forall \varphi \in M$ for any $n, m \in \mathbb{N}^*$. The assertion follows from Lemma 1.5.2.

Definition 1.5.5 Let K be a Hausdorff compact space and let M be a subset of continuous functions on K with values in the interval $[0, 1]$ which contains the constant functions 0 *and* 1. A subset $S \subset K$ is called antisymmetric with respect to M (M-antisymmetric) if any element $\varphi \in M$ with the property:

$$\varphi \cdot f + (1 - \varphi) \cdot g \, |S \in M \, |S, \forall f, g \in M,$$

is constant on S.

Furthermore, we denote by Σ the set of all antisymmetric subsets of K with respect to M. From Definition 1.5.5 we deduce the following remark.

Remark 1.5.6 For any $x \in K$ the set $\{x\}$ belongs to Σ, and there exists a maximal M-antisymmetric subset S_x such that $x \in S_x$. Moreover, for any $x, y \in K$ we have either $S_x \cap S_y = \phi$ or $S_x = S_y$ and $K = \cup \{S_x; \, x \in K\}$.

Theorem 1.5.7 *Let $M \subset C(K, \, [0, 1])$ be such that the constant functions $0, 1 \in M$ and let $f \in C(K, \, [0, 1])$. Then, there exists a subset $S \subset K$ antisymmetric with respect to M such that*

$$dist(f, M) = dist(f \, |S \, , M \, |S).$$

Proof We denote by $\mathcal{F}$ the family of all closed subsets $F \subset K$ with the property:

$$dist(f \, |F, M \, |F) = dist(f, M) = d.$$

By Lemma 1.4.5, $\mathcal{F}$ has minimal elements with respect to inclusion as order relation. To prove the theorem, it will be sufficient to show that a minimal element S of $\mathcal{F}$ is antisymmetric with respect to M. We suppose the contrary and let $\varphi \in M$ be such that

$$\varphi \cdot f + (1 - \varphi) \cdot g \, |S \in M \, |S, \forall f, g \in M,$$

but φ is not constant on S. We choose $y, z \in S$ and $a, b \in [0, 1]$ such that

$$0 \leq \varphi(y) < a < b < \varphi(z) \leq 1.$$

We may suppose that $2 \cdot a < b$. Indeed, since $\frac{a}{b} < 1$ we deduce that there exists $n \in \mathbb{N}^*$ such that

$$\left(\frac{a}{b}\right)^n < \frac{1}{2}, 2 \cdot a^n < b^n.$$

Obviously, we have

$$0 \leq \varphi^n(y) < a^n < b^n < \varphi^n(z) \leq 1.$$

Furthermore, we replace φ^n, a^n, b^n respectively by φ, a, b and so we have

$$\varphi \in M; 2 \cdot a < b; 0 \leq \varphi(y) < a < b < \varphi(z) \leq 1,$$

$$\frac{1}{a} > \frac{1}{a} - \frac{1}{b} \geq \frac{1}{2 \cdot a} > \frac{1}{b} > 1 \; ; \; 2 < \frac{2}{b} < \frac{1}{a}.$$

So there exists $k \in \mathbb{N}$ such that

$$\frac{1}{b} < k < \frac{1}{a} \; ; \; a < \frac{1}{k} < b.$$

Let now $p_n = (1 - \varphi^n)^{k^n}$. Then using Lemma 1.5.2 we get $p_n | S \in M | S$. Furthermore, we denote

$$Y = \{x \in S; 0 \leq \varphi(x) \leq b\}, Z = \{x \in S; a \leq \varphi(x) \leq 1\}.$$

Obviously, $Y \subset S$ and $Y \neq S$. Since S is minimal in $\mathcal{F}$ we get that $Y \notin \mathcal{F}$, and therefore there exists $m_Y \in M$ such that $\|f - m_Y\|_Y < d$. Similarly there exists $m_Z \in M$ such that $\|f - m_Z\|_Z < d$. We observe that if $x \in Y \backslash Z$ then $0 \leq \varphi(x) \leq a < \frac{1}{k}$. From Theorem 1.1.1, it follows that the sequence $(p_n)_n$ converges uniformly to 1 on $Y \backslash Z$, and similarly, if $x \in Z \backslash Y$ then $\frac{1}{k} < b \leq \varphi(x) \leq 1$, and therefore the sequence $(1 - p_n)_n$ converges uniformly to 1 on the set $Z \backslash Y$.

Now, if we denote by

$$h_n = p_n \cdot m_Y + (1 - p_n) \cdot m_Z,$$

we have

$$h_n \,|\, S \in M \,|\, S, \quad h_n \to m_Y \text{ on } Y \backslash Z, \quad h_n \to m_Z \text{ on } Z \backslash Y.$$

Hence for $n \in \mathbb{N}$ sufficiently large we have

$$\|f - h_n\|_{Y \backslash Z} < d, \|f - h_n\|_{Z \backslash Y} < d.$$

On the other hand if $x \in Y \cap Z$, then we have

$$|f(x) - h_n(x)|$$
$$= |p_n(x) \cdot f(x) + (1 - p_n(x)) \cdot f(x) - p_n(x) \cdot m_Y(x)$$
$$- (1 - p_n(x)) \cdot m_Z(x)| \le p_n(x) \cdot |f - m_Y|_Y + (1 - p_n(x)) \cdot |f - m_Z|_Z < d.$$

Since $S = (Y \backslash Z) \cup (Y \cap Z) \cup (Z \backslash Y)$, from the preceding considerations we arrive at the inequality $\|f - h_n\|_S < d$. Taking into account that $h_n \,|\, S \in M \,|\, S$, from the last inequality it follows that $S \notin \mathcal{F}$ and this is a contradiction.   □

Example 1.5.8 Let M be the set of continuous functions $p : [0, 1] \to [0, 1]$ with the properties:

$$p(x) = a_m \cdot x + b_m, \forall x \in J = \left[0, \frac{1}{2}\right], \ a_m, \ b_m \in \mathbb{R} \text{ and } p(1) \in \mathbb{Z}$$

We remark that

$$0, 1 \in M, \ S_x = S_0, \forall x \in J, \ S_x = \{x\}, \forall x \in \left(\frac{1}{2}, 1\right].$$

Indeed, if $p \in M$ has the property $p \cdot q + (1 - p) \cdot r \,|\, J \in M \,|\, J$, $\forall q, r \in M$, then taking $q = p$ and $r = 0$, it follows $p^2 \,|\, J \in M \,|\, J$, and therefore

$$a_m{}^2 \cdot x^2 + 2 \cdot a_m \cdot b_m \cdot x + b_m{}^2 = c \cdot x + d, \forall x \in J,$$

hence $a_m = 0$, and $p(x) = b_m$, $\forall x \in J$, i.e., J is an antisymmetric set with respect to M.

On the other hand, since any continuous function $\varphi : [0, 1] \to [0, 1]$ which is constant on $\left[0, \frac{1}{2}\right]$ and $\varphi(1.1) \in \{0, 1\}$ is such that M is φ-convex, we deduce that a subset S which contains at least an element $x \in \left(\frac{1}{2}, 1\right]$ can be antisymmetric with respect to M only in the case when this set is a singleton. So for any $x \in \left(\frac{1}{2}, 1\right]$ we have $S_x = \{x\}$.

Let now $f : [0, 1] \to [0, 1]$ be a continuous function. It is obvious that if $x \in \left(\frac{1}{2}, 1\right]$ we have

$$\{p(x); \ p \in M\} = [0, 1],$$

and therefore we have

$$dist\{f \,|S_x, \ M \,|S_x \} = 0.$$

From Theorem 1.5.7, we get

$$dist(f, M) = \max\{dist(f \,|J, M \,|J); \ dist(f(1), \{0, 1\}) \}.$$

As usual, for any $x \in K$ we denote

$$[x] = \{y \in K; m(y) = m(x), \forall m \in M\}.$$

Obviously, $[x]$ is the greatest constant subset for M, $x \in [x]$ and if $x, y \in K$ we have either $[x] = [y]$, or $[x] \cap [y] = \phi$. Hence the family $\{[x]; \ x \in K\}$ is a disjoint partition of K, i.e.,

$$K = \cup\{ [x] ; x \in K\}.$$

The following result was obtained by Prolla in [34].

Corollary 1.5.9 *Let $M \subset C(K, [0, 1])$ be a subset having (VN) property, containing the constant functions 0 and 1, and let $f : K \to [0, 1]$ be a continuous function. Then there exists $x \in K$ such that*

$$dist(f; \ M) = dist(f \,|\, [x] \,; \ M \,|\, [x]).$$

The assertion follows from Theorem 1.5.7, since in our case the family of antisymmetric subsets of K with respect to M coincides with the set $\{[x]; \ x \in K\}$.

Lemma 1.5.10 *For any $a, b, \varepsilon \in (0, 1)$ there exist $k, m \in \mathbb{N}^*$ such that*

$$\left|a - (1 - b^k)^m\right| < \varepsilon.$$

Proof We choose $k \in \mathbb{N}^*$ such that $b^k < \varepsilon$ and $a < 1 - b^k = d$. Also we consider $m \in \mathbb{N}^*$ such that $d^{m+1} < a \leq d^m$. We observe now that

$$d^m - d^{m+1} = d^m(1 - d) = d^m \cdot b^k \leq b^k < \varepsilon,$$

and therefore

$$\left| a - (1 - b^k)^m \right| = \left| a - d^m \right| < d^m - d^{m+1} < \varepsilon. \qquad \square$$

Theorem 1.5.11 (Prolla [34]) *Let $M \subset C(K, [0, 1])$ be a subset having (VN) property, containing the constant functions 0 and 1, and let $f \in C(K, [0, 1])$ be arbitrary. Then, we have $f \in \overline{M}$ iff the following conditions are fulfilled:*

a) For any $x, y \in K$ with the property $f(x) \neq f(y)$, there exists $m \in M$ such that

$$m(x) \neq m(y).$$

b) For any $x \in K$, with the property $0 < f(x) < 1$, there exists $m \in M$ such that

$$0 < m(x) < 1.$$

Proof Obviously, if $f \in \overline{M}$, the conditions $a)$ and $b)$ are satisfied.

Let $f \in C(K, [0, 1])$ be a function which satisfies the conditions $a)$ and $b)$. By Corollary 1.5.9, there exists $x_0 \in K$ such that

$$dist(f; M) = dist(f \mid [x_0]; \ M \mid [x_0]),$$

where $[x_0] = \{x \in K; \ m(x) = m(x_0), \ \forall m \in M\}$. Using now the condition $a)$ we get $f(x) = f(x_0)$ for all $x \in [x_0]$. If $f(x_0) = 0$, then $dist(f \mid [x_0]; \ M \mid [x_0]) = 0$ since the constant function 0 belongs to M. Similarly, if $f(x_0) = 1$, then $dist(f \mid [x_0]; \ M \mid [x_0]) = 0$ since the constant function 1 belongs to M. We analyse now the case $0 < f(x_0) < 1$.

Let $0 < \varepsilon < 1$ be arbitrary. From the condition $b)$ we may choose $m_0 \in M$ such that

$$0 < m_0(x_0) < 1.$$

By Lemma 1.5.10, there exist $k, p \in \mathbb{N}^*$ such that

$$\left| f(x_0) - [1 - m_0^{\ k}(x_0)]^p \right| < \varepsilon.$$

This inequality holds if we replace x_0 by any $x \in [x_0]$. If we denote

$$\varphi(x) = [1 - m_0{}^k(x_0)]^p, \; \forall x \in K,$$

then by Remark 1.5.4, we get $\varphi \in M$. Furthermore, we have

$$|f(x) - \varphi(x)| < \varepsilon, \; \forall x \in [x_0], \; dist\,(f \mid [x_0]; \; M \mid [x_0]\,) < \varepsilon. \qquad \square$$

Theorem 1.5.12 (Prolla [34]) *Let $M \subset C(K, \; [0, 1])$ be a closed subset with respect to the uniform topology on $C(K, \; [0, 1])$, having VN property and containing the constant functions 0 and 1. Then, M is a lattice with respect to the pointwise order relation.*

Proof Using Theorem 1.5.11, we show that

$$f = m_1 \vee m_2 = \sup\{m_1, m_2\} \in M, \; \forall m_1, m_2 \in M.$$

For this we remark that f satisfies the condition $b)$ from Theorem 1.5.11. Now if $x, y \in K$ are such that $f(x) \neq f(y)$ then one of the relations $m_1(x) = m_1(y), \; m_2(x) = m_2(y)$ fails, and therefore the condition $b)$ in the same theorem is fulfilled. In a similar way, one can show that the function $g = m_1 \wedge m_2 = \inf\{m_1, m_2\} \in M, \; \forall m_1, m_2 \in M$. Hence $f, g \in \overline{M} = M$. $\qquad \square$

Theorem 1.5.13 *Let $M \subset C(K, \; [0, 1])$ be a subset with VN property, containing the constant functions 0 and 1, separates the points of K, and for any $x \in K$ there exists $m \in M$ such that $0 < m(x) < 1$. Then M is dense in $C(K, \; [0, 1])$.*

Proof Indeed, the assertion may be deducted directly from Theorem 1.5.11 since any function $f \in C(K, \; [0, 1])$ satisfies the conditions $a)$ and $b)$ of this theorem. $\qquad \square$

Corollary 1.5.14 *Let $P_1{}^+(I^n)$ be the set of all polynomial functions $p : I^n \to I = [0, 1]$, where I^n is the unit cube in $\mathbb{R}^n$, i.e., $I^n = \underbrace{I \times I \times \ldots \times I}_{n}$. Then $P_1{}^+(I^n)$ is dense in $C(I^n, \; [0, 1])$.*

Proof The assertion follows from Theorem 1.5.13 since the set $M = P_1{}^+(I^n)$ satisfies the hypotheses of this theorem. $\qquad \square$

Theorem 1.5.15 (Jewett [12]) *Let $a, b \in \mathbb{R}, \; 0 \leq a < b \leq 1$. Then, for any $\varepsilon > 0$, there exist $m, n \in \mathbb{N}^*$ such that the polynomial function $p(x) = (1 - x^m)^n, \; x \in [0, 1]$ satisfies* *the following inequalities:*

$$(i) \ p(x) > 1 - \varepsilon, \ \forall x \in [0, a],$$

$$(ii) \ p(x) < \varepsilon, \ \forall x \in [b, 1].$$

Proof There is no loss of generality if we suppose $0 < a < b < 1$. From Bernoulli inequality, we have

$$\left(1 - x^m\right)^n \geq \left(1 - a^m\right)^n \geq 1 - n \cdot a^m, \ \forall x \in [0, a] \quad \text{and}$$

$$\left(1 - x^m\right)^n \leq \left(1 - b^m\right)^n \leq \frac{1}{(1 + b^m)^n} \leq \frac{1}{1 + n \cdot b^m} \leq \frac{1}{n \cdot b^m}, \ \forall x \subset [b, 1].$$

We remark that we may choose $m, n \in \mathbb{N}^*$ such that

$$1 - n \cdot a^m \geq 1 - \varepsilon \text{ and } \frac{1}{n \cdot b^m} < \varepsilon,$$

or equivalently

$$\frac{1}{\varepsilon \cdot b^m} < n < \frac{\varepsilon}{a^m}.$$

The last two inequalities hold if we take $m \in \mathbb{N}^*$ such that

$$\frac{2}{\varepsilon \cdot b^m} < \frac{\varepsilon}{a^m} \text{ or } \frac{2}{\varepsilon^2} < \left(\frac{b}{a}\right)^m,$$

and we take

$$n = 2 \cdot \left[\frac{1}{\varepsilon \cdot b^m} \right]. \qquad \qquad \square$$

Lemma 1.5.16 *Let $M \subset C(K, [0, 1])$ be a subset with VN property, containing the constant functions 0 and 1 and separating the points of K. Let $x_0 \in K$ and let U be an open neighbourhood of x_0. Then there exists an open neighborhood V of x_0, $V \subset U$ such that for any $\varepsilon \in (0, 1)$ there exists $m \in \overline{M}$ such that*

1) $m(x) < \varepsilon$ *if $x \notin U$,*
2) $m(x) > 1 - \varepsilon$ *if $x \in V$.*

Proof We denote by F the compact set $K \backslash U$. If $y \in F$, then $y \neq x_0$ and since M separates the points of K, there exists $m_y \in M$ such that $m_y(y) \neq m_y(x_0)$, and let us suppose

$$m_y(y) < m_y(x_0).$$

From Theorem 1.5.15, we deduce that the existence of $m, n \in \mathbb{N}^*$ such that the polynomial function $p_y(x) = (1 - x^m)^n$ has the properties:

$$p_y[m_y(y)] > \frac{3}{4}, \text{ and } p_y[m_y(x_0)] < \frac{1}{4}.$$

By Remark 1.5.4, we get $p_y(m_y) \in M$, and we denote

$$W_y = \left\{ x \in K; \ p_y[m_y(x)] > \frac{3}{4} \right\},$$

then $x_0 \notin W_y$ and $y \in W_y$. Since F is a compact set, we may choose a finite system of points $y_i \in F$, $i = 1, 2, \ldots, k$ such that

$$F \subset \bigcup_{i=1}^{k} W_{y_i}.$$

The function $h = g_1 \vee \ldots \vee g_k$, where $g_i = p_{y_i}(m_{y_i})$, $i = \overline{1, k}$, belongs to $\overline{M}$ according to Theorem 1.5.12. We observe that

$$h(x_0) < \frac{1}{4} \text{ and } h(x) > \frac{3}{4}, \forall x \in F.$$

Let $V = \left\{ x \in K; \ h(x) < \frac{1}{4} \right\}$ and let $q_n = (1 - h^n)^{3^n}$. We remark that $x_0 \in V \subset U$ and $q_n \in \overline{M}$.

On the other hand, from Theorem 1.1.1, where $a = \frac{1}{4} < \frac{1}{k} = \frac{1}{3} < b = \frac{3}{4}$, we deduce that the sequence $\{q_n\}_n$ converges uniformly to the constant function 1 on V and to the constant function 0 on F. Hence for a sufficiently large n the function $m = q_n$ satisfies the inequalities:

$$m > 1 - \varepsilon \text{ on } V \text{ and } m < \varepsilon \text{ on } F = K \backslash U. \qquad \square$$

Theorem 1.5.17 *Let $M \subset C(K, [0, 1])$ be a subset which VN property, containing the constant functions 0 and 1 and separating the points of K. Then, $\overline{M}$ is an Uryson family on K, i.e., if A and B are two disjoint closed subsets of K and $0 < \varepsilon < 1$, then there* *exists $m \in \overline{M}$ such that*

$$m(x) < \varepsilon, \ \forall x \in A, \ \ m(x) > 1 - \varepsilon, \ \forall x \in B.$$

Proof If we denote $U = K \backslash B$, then U is an open neighbourhood of A. From Lemma 1.5.16, for any $x \in A$ there exists an open neighbourhood V_x of x, $V_x \subset U$ and there exists a function $m_x \in \overline{M}$ such that

$$m_x < \varepsilon \ \text{on} \ _x \ \text{and} \ m_x > 1 - \varepsilon \ \text{on} \ B.$$

Since A is compact and $A \subset \bigcup\limits_{x \in A} V_x$, it follows that there exists $x_1, \ldots, x_n \in A$ such that

$$A \subset \bigcup_{i=1}^{n} V_{x_i}.$$

Using again Lemma 1.5.16, for any $i = \overline{1, n}$ there exists $m_i \in \overline{M}$ with the properties:

$$m_i(x) < \frac{\varepsilon}{n}, \forall x \in V_{x_i}, \ \text{and} \ m_i(x) > 1 - \frac{\varepsilon}{n}, \forall x \in B.$$

If we denote $m(x) = m_1(x) \cdot m_2(x) \cdot \ldots \cdot m_n(x)$, $\forall x \in K$, then $m \in \overline{M}$, and for any $x \in A$ there exists $1 \leq j \leq n$ such that $x \in V_{x_j}$, i.e., $m(x) < \frac{\varepsilon}{n} < \varepsilon$. If $x \in B$ then $m(x) > \left(1 - \frac{\varepsilon}{n}\right)^n \geq 1 - n \cdot \frac{\varepsilon}{n} = 1 - \varepsilon$. $\qquad \square$

Remark 1.5.18 Let K be a Hausdorff compact space and let $\mathcal{A}$ be an algebra of continuous real functions on K which contains the constant functions and separates the points of K. Then, the set $\mathcal{A}_1 = \{a \in \mathcal{A}; \ 0 \leq a \leq 1\}$ is an Uryson family on K.

The assertion follows immediately from Theorem 1.5.17.

Theorem 1.5.19 *Let $M \subset C(K, \ [0, 1])$ be a uniformly closed subset having VN property, containing the constant functions 0 and 1 and separating the points of K. Moreover we suppose that M contains a constant function c, $0 < c < 1$. Then*

$$M = C(K, \ [0, 1]).$$

Proof Using Theorem 1.5.17, it follows that M is an Uryson family on K. On the other hand, from Lemma 1.5.10, we deduce that any constant function λ, $\lambda \in (0, 1)$ belongs to M, hence M is a convex set. From Theorem 1.3.6, we get $M = C(K, \ [0, 1])$. $\qquad \square$

Approximation of Continuous Functions on Locally Compact Spaces

2.1   Weighted Spaces of Scalar Functions

Throughout this section X will be a locally compact Hausdorff space, and $\mathbb{K}$ the real or complex numbers. Also, we shall denote by $C(X, \mathbb{K}$, or simply $C(X)$, the space of all continuous functions on X with values in $\mathbb{K}$.

Definition 2.1.1 A family V of upper semicontinuous, non-negative functions on X such that for any $v_1, v_2 \in V$ and any $\lambda \in R$, $\lambda > 0$ there exists $w \in V$ such that

$$v_i(x) \leq \lambda \cdot w(x), \forall x \in X, i = 1, 2,$$

will be called a Nachbin family on X. Any element of V will be called a weight.

We shall denote by $CV_b(X, \mathrm{K})$ or by $CV_b(X)$ the set of all continuous functions f on X such that the function $f \cdot v$ is bounded on X for all $v \in V$. Any weight $v \in V$ generates a seminorm $p_v : CV_b(X) \to \mathbb{R}_+$ defined by

$$p_v(f) = \sup\{v(x) \cdot |f(x)| ;\ x \in X\}, \quad \forall f \in CV_b(X).$$

The locally convex topology defined by this family of seminorms is denoted by ω_V and it will be called the weighted topology on $CV_b(X)$.

The family of seminorms $(p_v)_{v \in V}$ is upper directed, and the family $(B_v)_{v \in V}$ of subsets of $CV_b(X)$ is a base of neighbourhoods of the origin, where

$$B_v = \{f \in CV_b(X) ;\ p_v(f) \leq 1\}.$$

© The Editor(s) (if applicable) and The Author(s), under exclusive licence
to Springer Nature Switzerland AG 2020
I. Bucur, G. Paltineanu, *Topics in Uniform Approximation of Continuous Functions*,
Frontiers in Mathematics, https://doi.org/10.1007/978-3-030-48412-5_2

One can see that for any continuous bounded function $g : X \to K$ we have $g \cdot f \in CV_b(X)$ for any $f \in CV_b(X)$, and moreover

$$p_v(g \cdot f) \le \|g\| \cdot p_v(f),$$

where $\|g\|$ is the uniform norm of g.

Also, we remark that $CV_b(X)$ is an order ideal in the $C(X)$ endowed with the pointwise order relation; more precisely if $f \in CV_b(X)$ and $g \in C(X)$ are such that $|g| \le |f|$, then $g \in CV_b(X)$ and $p_v(g) \le p_v(f)$, $\forall v \in V$.

As for uniqueness of the limit in the locally convex space $(CV_b(X), \omega_V)$, one can see that this space is Hausdorff iff the interior of the set:

$$\bigcap_{v \in V} [v = 0] = \bigcap_{v \in V} \{x \in X \; ; \; v(x) = 0\}$$

in the topological space X is empty.

In order to use measure theory methods in our approach, we shall restrict to an order ideal of the space $CV_b(X)$ denoted by $CV_0(X)$, the space of all functions $f \in C(X)$ such that the function $f \cdot v$ vanishes at infinity, i.e., the space of all functions $f \in C(X)$ having the property that for any $v \in V$ and any $\varepsilon > 0$ the set:

$$[v \cdot |f| \ge \varepsilon] = \{x \in X; v(x) \cdot |f(x)| \ge \varepsilon\}$$

is a compact subset of the starting topological space X.

We endow the space $CV_0(X)$ with the trace of the above topology ω_V and we call it the weighted space associated with the Nachbin family V.

Remark 2.1.2

(a) If for any point $x \in X$ there exists $v_x \in V$ such that $v_x(x) > 0$, then $CV_b(X)$ in particular $CV_0(X)$ are locally convex Hausdorf spaces.

(b) If for any point $x \in X$ there exists $v_x \in V$ and there exists $r \in R$, $r > 0$ such that the set:

$$[v_x > r] = \{y \in X; v_x(y) > r\},$$

is a neighbourhood of x in X, then the locally convex spaces $(CV_b(X), \omega_V)$ and $(CV_0(X), \omega_V)$ are complete spaces.

Remark 2.1.3 The weighted space $CV_0(X)$ is a closed subspace of $CV_b(X)$.

Indeed, let $g \in \overline{CV_0(X)}$. Then, for any $\varepsilon > 0$ and any $v \in V$ there exists $f \in CV_0(X)$ such that

$$p_v(g - f) < \frac{\varepsilon}{2}.$$

Since $f \in CV_0(X)$, the set $\left[v \cdot |f| \geq \frac{\varepsilon}{2}\right] = \left\{x \in X; v(x) \cdot |f(x)| \geq \frac{\varepsilon}{2}\right\}$ is compact. On the other hand,

$$v(x) \cdot |g(x)| \leq v(x) \cdot |g(x) - f(x)| + v(x) \cdot |f(x)| < \frac{\varepsilon}{2} + v(x) \cdot |f(x)|, \forall x \in X,$$

and therefore we have

$$[v \cdot |g| \geq \varepsilon] \subset \left[v \cdot |f| \geq \frac{\varepsilon}{2}\right].$$

It follows that the set $[v \cdot |g| \geq \varepsilon]$ is compact, hence $g \in CV_0(X)$.

In the sequel, we shall denote by $\mathcal{K}_{\mathbb{C}}(X)$ (resp. $\mathcal{K}_{\mathbb{R}}(X)$) the set of all continuous functions on X with complex values (resp. real values) and having compact support. Sometimes, we use the notation $\mathcal{K}(X)$ for $\mathcal{K}_{\mathbb{C}}(X)$ as well as $\mathcal{K}_{\mathbb{R}}(X)$.

Proposition 2.1.4 *Let X be a locally compact Hausdorff space and let V be a Nachbin family on X. Then, we have $\mathcal{K}(X) \subset CV_b(X)$ and $\mathcal{K}(X)$ is dense in $CV_0(X)$, i.e. $\overline{\mathcal{K}(X)} = CV_0(X)$.*

Proof Since for any element $\varphi \in \mathcal{K}(X)$ and any $v \in V$ the function $\varphi \cdot v$ vanishes outside the support of φ, it follows that $\varphi \in CV_0(X)$, hence $\mathcal{K}(X) \subset CV_0(X) \subset CV_b(X)$.

Let now $f \in CV_0(X)$, $\varepsilon > 0$ and $v \in V$ be arbitrarily chosen. If we denote by $Y = [|f| \cdot v \geq \varepsilon]$, then Y is a compact set, and therefore there exists a function $\varphi \in \mathcal{K}(X)$, $\varphi : X \to [0, 1]$ such that $\varphi(x) = 1, \forall x \in Y$. Furthermore, we have $\varphi \cdot f \in \mathcal{K}(X)$, and

$$p_v(f - f \cdot \varphi) = \sup_{x \in X} |(1 - \varphi)f \cdot v)| (x) = \sup_{x \in X \setminus Y} [1 - \varphi(x)] \cdot |f(x)| \cdot v(x) < \varepsilon. \qquad \square$$

Proposition 2.1.5 *Let $f \in CV_0(X)$ and $v \in V$ be arbitrarily chosen. Then, the restriction of the function $v \cdot |f|$ to any closed subset F of X has a maximum on this set.*

Proof Indeed, if there exists $x_0 \in F$ such that $v(x_0) \cdot |f(x_0)| = \varepsilon > 0$, then the subset $\left[v \cdot |f| \geq \frac{\varepsilon}{2}\right]$ is compact, as well as the set $F \cap \left[v \cdot |f| \geq \frac{\varepsilon}{2}\right]$.

Since the function $v \cdot |f|$ is upper semicontinuous, it follows that there exists $y_0 \in F \cap \left[v \cdot |f| \geq \frac{\varepsilon}{2}\right]$ such that

$$(v \cdot |f|) (y_0) = \sup \{v(x) \cdot |f(x)|; x \in F\}. \qquad \square$$

Example 2.1.6 Let X be a Hausdorff locally compact space and let V be the family of all characteristic functions of compact subsets of X. In this case, we have
$CV_b(X) = CV_0(X) = C(X)$ (the space of all continuous functions on X) and the weighted topology ω_v coincides with the topology of uniform convergence on any compact subset of X.

Example 2.1.7 Let X be a Hausdorff locally compact space. If the Nachbin family V reduces to the constant function 1, i.e. $V = \{\,1\,\}$, then $CV_b(X) = C_b(X)$—the space of all continuous and bounded functions on X, and ω_v coincides with the topology of uniform convergence on $C_b(X)$, i.e., the topology given by the norm:

$$\|f\| = \sup\{\,|f(x)|;\ x \in X\},\ \forall f \in C_b(X).$$

In this case, $CV_0(X) = C_0(X)$—the space of all continuous functions on X which vanish at infinity.

Example 2.1.8 Let X be a Hausdorff locally compact space, and let $V = C_0^+(X)$ be the family of all non-negative continuous functions on X vanishing at infinity. Then, $CV_0(X) = C_b(X)$ and the weighted topology ω_v coincides with the strict topology β on $C_b(X)$.

We remember that this notion and notation was introduced by R.C.Buck (see [6]) and is given by the family of seminorms:

$$p_v(f) = \sup\{v(x) \cdot |f(x)|\,;\ x \in X\},\ \forall v \in C_0^+(X), \forall f \in C_b(X).$$

Since the inclusion $C_b(X) \subset CV_0(X)$ is obvious, we show only the inclusion $CV_0(X) \subset C_b(X)$.

Indeed, if $f \in C^+(X)\backslash C_b(X)$, then there exists a strictly increasing and unbounded sequence of positive numbers $(a_n)_n$ such that the open sets:

$$[a_n < f < a_{n+1}] = \{x \in X;\ a_n < f(x) < a_{n+1}\}$$

are nonempty. For any $n \in \mathbb{N}$ we consider a point $x_n \in [a_n < f < a_{n+1}]$, a compact neighbourhood K_n of x_n, $K_n \subset [a_n < f < a_{n+1}]$, and a function $\varphi_n \in \mathcal{K}(X)$ such that

$$0 \leq \varphi_n \leq 1,\quad \varphi_n(x_n) = 1,\quad \varphi_n = 0\ \text{on}\ X\backslash K_n.$$

Obviously, the function $\varphi = \sum_{i=1}^{\infty} \frac{1}{\sqrt{a_n}} \cdot \varphi_n$ belongs to $C_0^+(X)$ because $\lim\limits_{n\to\infty} \frac{1}{\sqrt{a_n}} = 0$. On the other hand, since $f(x_n) \cdot \varphi(x_n) > \sqrt{a_n}$, we deduce that the continuous function $f \cdot \varphi$ is unbounded, and therefore it does not vanish at infinity, i.e, $f \notin CV_0(X)$.

Example 2.1.9 Let $X = \mathbb{R}^n$, and let $\mathscr{P}_n$ be the set of all polynomials defined on $\mathbb{R}^n$ with values in K. If $V = \{ \ |p| \ ; \forall p \in \mathscr{P}_n\}$, then $CV_b(\mathbb{R}^n) = CV_0(\mathbb{R}^n) \subset C_b(\mathbb{R}^n)$. The elements of $CV_b(\mathbb{R}^n)$ are called rapidly decreasing at infinity.

Indeed, if we take an element $f \in CV_b(\mathbb{R}^n)$ and $p \in \mathscr{P}_n$, then the function:

$$x \to \left(1 + \|x\|^2\right) \cdot p(x) \cdot f(x) : \mathbb{R}^n \to \text{K}, \quad \left(\|x\|^2 = \sum_{i=1}^{n} x_i{}^2\right),$$

is bounded on $\mathbb{R}^n$ and therefore, for some $M \in \mathbb{R}_+$ we have

$$|p(x) \cdot f(x)| \leq \frac{M}{1 + \|x\|^2},$$

hence $\lim_{x \to \infty} p(x) \cdot f(x) = 0, \quad \forall f \in CV_b(\mathbb{R}^n)$.

It is not difficult to show that $CV_0(\mathbb{R}^n) = CW_0(\mathbb{R}^n)$ where

$$W = \left\{w_k; w_k : \mathbb{R}^n \to \mathbb{R}_+, \ k \in \mathbb{N}\right\}, \ w_k(x) = (1 + \|x\|)^k.$$

2.2   Duality for Weighted Spaces

In this part, with X being a Hausdorff locally compact space, $\mathcal{K}(X)$ will be the space of all continuous real or complex functions on X with compact support. If K is a compact subset of X, we shall denote by $\mathcal{K}(X, K)$ the set of all continuous scalar functions on X which vanishes outside K.

Obviously, if we endow $\mathcal{K}(X)$ with uniform norm topology, then any $\mathcal{K}(X, K)$ is a complete vector subspace of $\mathcal{K}(X)$, namely for any compact K of X, $\mathcal{K}(X, K)$ coincides with the set of all continuous scalar functions defined on K and vanishes at the boundary of K in X. Nevertheless, the vector space:

$$\mathcal{K}(X) = \bigcup_{K} \mathcal{K}(X, K)$$

endowed with the uniform norm is not a Banach space. The closure of $\mathcal{K}(X)$ in the space $C_b(X)$ of all continuous and bounded scalar functions on X is just the space $C_0(X)$ of all continuous scalar functions on X vanishing at infinity. This is why the normed spaces $C_0(X)$ and $\mathcal{K}(X)$ have the same dual.

Besides, the topology of uniform convergence on $\mathcal{K}(X)$ is considered the inductive limit topology denoted by τ_{ind}, i.e., the finest locally convex topology for which the injection maps:

$$i_K : \mathcal{K}(X, K) \to \mathcal{K}(X)$$

are continuous when K runs the sets of all compact subsets of X, and any vector space $\mathcal{K}(X, K)$ is endowed with the topology of uniform convergence. What is important here is the fact that giving an arbitrary locally convex space E and a linear map:

$$T : (\mathcal{K}(X), \tau_{ind}) \to E$$

then T is continuous iff the restriction of T to any $\mathcal{K}(X, K)$ is continuous.

Lemma 2.2.1 *If the Nachbin family V is the set of all non-negative continuous functions on X, i.e., $V = C^+(X)$, then*

$$\mathcal{K}(X) \subset CV_0(X) \subset C_0(X) \subset C_b(X),$$

the weighted topology ω_V on $CV_0(X)$ is finer than the topology of uniform convergence, and the trace on $\mathcal{K}(X)$ of ω_V is smaller than τ_{ind}.

Proof The first inclusion is valid for all family $V \subset C^+(X)$, and the second follows from the inclusion $C_b{}^+(X) \subset V = C^+(X)$ and Example 2.1.7. The weighted topology ω_V on $CV_0(X)$ is given by the seminorms:

$$p_v(f) = \sup \{v(x) \cdot |f(x)| \, ; \, x \in X\} , \, v \in V = C^+(X), \forall f \in CV_0(X),$$

whereas the topology of uniform convergence is given by one of these seminorms, namely that corresponding to $v = 1$, the constant function 1 on X.

Let us take now an arbitrary element $v \in V = C^+(X)$ and an arbitrary compact subset $K \subset X$. If we denote $\alpha = \sup \{ v(x); x \in K\}$, then $\alpha \in \mathbb{R}_+$ and the trace on $\mathcal{K}(X, K)$ of the neighbourhood B_v of the origin of $CV_0(X)$, where $B_v = \{f \in CV_0(X); \, p_v(f) \leq 1\}$ and contains all elements $f \in \mathcal{K}(X, K)$ for which

$$\|f\| = \sup \{|f(x)| \, ; \, x \in K\} < \frac{1}{\alpha}.$$

Hence the natural injection $i_K : \mathcal{K}(X, K) \to CV_0(X)$, $i_K(f) = f$ is continuous, and therefore the topology ω_V is smaller than the topology τ_{ind} on $\mathcal{K}(X, K)$. $\qquad\square$

Lemma 2.2.2 *If the locally compact space X is countable at infinity, then we have $\mathcal{K}(X) = CV_0(X)$ and $\tau_u \subset \omega_V \subset \tau_{ind}$, where $V = C^+(X)$.*

Proof Let $(K_n)_n$ be an increasing sequence of compact subsets of X such that

$$X = \bigcup_{n=0}^{\infty} K_n, \quad K_m \subset \mathring{K}_{m+1}, \quad \forall m \in \mathbb{N}.$$

We know that $CV_0(X) \subset C_0(X)$, and for any function $f \in C_0(X) \backslash \mathcal{K}(X)$ there exists a strictly increasing sequence $(n_p)_p$ of natural numbers, a strictly decreasing sequence $(\alpha_p)_p$ of real numbers, and there exists a sequence $(x_p)_{p \geq 1}$ in X such that

$$x_p \in \overset{\circ}{K}_{n_p} \backslash K_{n_{p-1}}, \ f(x_p) = \alpha_p, \ \lim_{p \to \infty} \alpha_p = 0.$$

For any $p \geq 1$, we consider a continuous function φ_p on X such that

$$0 \leq \varphi_p \leq 1, \ \varphi_p(x_p) = 1, \ \varphi_p = 0 \ \text{on} \ X \backslash \overset{\circ}{K}_{n_p}.$$

The function $\varphi : X \to [0, \infty)$ given by

$$\varphi(x) = \sum_{p=1}^{\infty} \frac{1}{a_p{}^2} \cdot \varphi_p(x)$$

is continuous, and therefore $\varphi \in V$. Since $\varphi(x_p) \cdot f(x_p) = \frac{1}{\alpha_p}$ for any $p \in \mathbb{N}$, we deduce that $\varphi \cdot f \notin C_0(X)$. Hence $f \notin CV_0(X)$, and therefore $CV_0(X) \subset \mathcal{K}(X)$.

By definition, a Radon measure on X is a linear scalar function m defined on $\mathcal{K}(X)$ which is continuous with respect to the topology τ_{ind}. This means that for any compact subset $K \subset X$ there exists a real number $a_K > 0$ such that

$$|m(f)| \leq a_K \cdot \|f\|_K = a_K \cdot \sup\{|f(x)|; \ x \in K\}, \ \forall f \in \mathcal{K}(X, K).$$

One can show that for any Radon measure m on X there exists a measure $|m| : \mathcal{K}(X) \to \mathbb{R}$ such that for any $f \in \mathcal{K}^+(X)$ we have

$$|m|(f) = \sup\{|m(g)|; \ g \in \mathcal{K}(X), \ |g| \leq f\},$$

and the following inequality holds:

$$|m(f)| \leq |m|(|f|), \ \forall f \in \mathcal{K}(X).$$

In fact, $|m|$ is the smallest positive measure on $\mathcal{K}(X)$ satisfying the previous inequality. It is known that for any real measure $\mu : \mathcal{K}(X) \to \mathbb{R}$ there exist two positive measures μ^+, μ^- such that

$$\mu = \mu^+ - \mu^-, \ \mu^+ = \frac{1}{2} \cdot (|\mu| + \mu), \ \mu^- = \frac{1}{2} \cdot (|\mu| - \mu).$$

We shall denote by $\mathcal{M}_b(X)$ the set of all bounded measure on X, i.e., the dual of the locally convex space $\mathcal{K}(X)$ endowed with the topology of uniform norm. Hence a Radon measure

m on X belongs to $\mathcal{M}_b(X)$ iff there exists a positive number α_m such that

$$|m(f)| \le \alpha_m \cdot \|f\|, \quad \forall f \in \mathcal{K}(X).$$

Any positive Radon measure μ on X may be extended to C_i—the set of all lower semicontinuous functions $\varphi : X \to (-\infty, \infty]$ with the property:

$$[\varphi < 0] = \{x \in X; \ \varphi(x) < 0\}$$

is relatively compact.

If we endow C_i with pointwise addition, pointwise order relation and natural multiplication with positive numbers, then C_i is a min-stable ordered convex cone such that the pointwise supremum of any subset of C_i belongs to C_i. For any $\varphi \in C_i$ we have

$$\varphi = \sup \{f \in \mathcal{K}(X); \ f \le \varphi\}.$$

For any $\varphi \in C_i$, we shall denote by $\bar{\mu}(\varphi)$ the element from $(-\infty, \infty]$ given by

$$\bar{\mu}(\varphi) = \sup \{\mu(f); \ f \in \mathcal{K}(X), \ f \le \varphi\}.$$

One can show that

(a) $\bar{\mu}(\alpha \cdot \varphi' + \beta \cdot \varphi'') = \alpha \cdot \bar{\mu}(\varphi') + \beta \cdot \bar{\mu}(\varphi''), \forall \alpha, \beta \in \mathbb{R}_+, \ \forall \varphi', \varphi'' \in C_i$
(b) $\varphi' \le \varphi'' \Rightarrow \bar{\mu}(\varphi') \le \bar{\mu}(\varphi'')$
(c) $\bar{\mu} \left(\sup_{i \in I} \varphi_i \right) = \sup_{i \in I} \bar{\mu}(\varphi_i),$

for any upper directed subset $(\varphi_i)_{i \in I}$ of C_i.

For any function $g : X \to [-\infty, \infty]$, we denote by $\mu^*(g)$ the element of $[-\infty, \infty]$ given by

$$\mu^*(g) = \inf \{\overline{\mu}(\varphi); \ \varphi \in C_i, \ \varphi \ge g\}.$$

The following properties may be shown:

(a) $\mu^*(f + g) \le \mu^*(f \wedge g) + \mu^*(f \vee g) \le \mu^*(f) + \mu^*(g)$, whenever the algebraic operations make sense.
(b) $\mu^*(\alpha \cdot f) = \alpha \cdot \mu^*(f), \forall \alpha \in \mathbb{R}_+$ and $\mu^*(f) \le \mu^*(g)$ if $f \le g$.
(c) For any increasing sequence $(f_n)_n$ of functions $f_n : X \to [-\infty, \infty]$ such that $\mu^*(f_1) > -\infty$, we have

$$\mu^* \left(\bigvee_n f_n \right) = \bigvee_n \mu^*(f_n)$$

where $\wedge$ (resp. $\vee$) means the pointwise infimum (resp. supremum).

We also extend μ from $\mathcal{K}(X)$ to C_s, where C_s is the set of upper semicontinuous functions, $\psi : X \to [-\infty, \infty)$ with the property $[\psi > 0] = \{x \in X; \; \psi(x) > 0\}$ is relatively compact.

If we endow C_s with the pointwise order relation and algebraic operations, then C_s is a max-stable ordered convex cone, and the pointwise infimum of any subset of C_s belongs to C_s.

In fact, for any $\psi \in C_s$ we have

$$\psi = \inf\{f \in \mathcal{K}(X); \; f \geq \psi\}.$$

For any $\psi \in C_s$, we denote by $\underline{\mu}(\psi)$ the element from $[-\infty, \infty)$ given by

$$\underline{\mu}(\psi) = \inf\{\mu(f); \; f \in \mathcal{K}(X), \; f \geq \psi\}.$$

One can show that

(a) $\underline{\mu}(\alpha \cdot \varphi' + \beta \cdot \varphi'') = \alpha \cdot \underline{\mu}(\varphi') + \beta \cdot \underline{\mu}(\varphi'')$, $\forall \alpha, \beta \in \mathrm{R}^*_+$, $\forall \varphi', \varphi'' \in C_s$,
(b) If $\varphi' \leq \varphi''$, then $\underline{\mu}(\varphi') \leq \underline{\mu}(\varphi'')$,

(c) $\underline{\mu}\left(\inf_{i \in I} \psi_i\right) = \inf_{i \in I} \underline{\mu}(\psi_i)$ for any lower directed subset $(\psi_i)_{i \in I}$ of C_s.

We remark that $C_s = -C_i$ and $\underline{\mu}(\psi) = -\overline{\mu}(-\psi)$ for any $\psi \in C_s$.

For any function $g : X \to [-\infty, \infty]$, we denote by $\mu_*(g)$ the element of $[-\infty, \infty]$ given by

$$\mu_*(g) = \sup\left\{\underline{\mu}(\psi); \; \psi \in C_s, \; \psi \leq g\right\}.$$

The following properties may be shown:

(a) $\mu_*(f + g) \leq \mu_*(f \wedge g) + \mu_*(f \vee g) \leq \mu_*(f) + \mu_*(g)$, whenever the algebraic operations make sense.
(b) $\mu_*(\alpha \cdot f) = \alpha \cdot \mu_*(f)$, $\forall \alpha \in \mathrm{R}^*_+$ and $\mu_*(f) \leq \mu_*(g)$ if $f \leq g$.
(c) For any decreasing sequence $(f_n)_n$ of functions $f_n : X \to [-\infty, \infty]$

such that $\mu_*(f_1) < \infty$, we have

$$\mu_*\left(\bigwedge_n f_n\right) = \bigwedge_n \mu_*(f_n).$$

From the above considerations, we deduce $\mu_*(f) = -\mu^*(-f)$ and $\mu_*(f) \leq \mu^*(f)$ for any function $f : X \to [-\infty, \infty]$. Also $\mu_*(f) = \mu^*(f)$ for any $f \in C_i \cup C_s$.

By definition, a function $f : X \rightarrow [-\infty, \infty]$ is μ–integrable if we have $\mu_*(f) = \mu^*(f) \in \mathbb{R}$. In this case, we denote simply by $\mu(f)$ the common values $\mu_*(f) = \mu^*(f)$.

The set of all μ–integrable functions will be denoted by $\mathcal{L}^1(\mu)$. Using the stated assertions for μ_* and μ^*, we deduce the following properties for $\mathcal{L}^1(\mu)$ and for the map $\mu : \mathcal{L}^1(\mu) \rightarrow \mathbb{R}$:

(a) For any $f, g \in \mathcal{L}^1(\mu)$, the functions $f \wedge g$ and $f \vee g$ belong to $\mathcal{L}^1(\mu)$.

(b) For any increasing sequence $(f_n)_n$ of $\mathcal{L}^1(\mu)$ such that the sequence $(\mu(f_n))_n$ is bounded, we have $\bigvee_n f_n \in \mathcal{L}^1(\mu)$ and $\mu(\bigvee_n f_n) = \sup_n \mu(f_n) = \lim_{n \to \infty} \mu(f_n)$.

(c) If $f, g \in \mathcal{L}^1(\mu)$ are finite, and $\alpha, \beta \in \mathbb{R}$, then we have

$$\alpha \cdot f + \beta \cdot g \in \mathcal{L}^1(\mu) \text{ and } \mu(\alpha \cdot f + \beta \cdot g) = \alpha \cdot \mu(f) + \beta \cdot \mu(g).$$

(d) If $(f_n)_n$ is a sequence of $\mathcal{L}^1(\mu)$ which converges pointwise to a function f, and there exists a function $g \in \mathcal{L}^1(\mu)$ such that $|f| \le g$, $\forall n \in \mathbb{N}$, then $f \in \mathcal{L}^1(\mu)$, $\mu(f) = \lim_{n \to \infty} \mu(f_n)$ and

$$\lim_{n \to \infty} \mu(|f - f_n|) = 0.$$

(e) If $f \in C_i \cup C_s$, then $f \in \mathcal{L}^1(\mu)$ iff $\mu(f) \in \mathbb{R}$.

(f) $\mathcal{K}(X) \subset \mathcal{L}^1(\mu)$.

We have extended the starting Radon measure $\mu : \mathcal{K}(X) \rightarrow \mathbb{R}$ to the set $\mathcal{L}^1(\mu)$ which contains any continuous function $f : X \rightarrow \mathbb{R}_+$ such that $\mu^*(f) < \infty$.

On the set $\mathscr{P}(X)$ of all subsets of X we consider the map:

$$A \rightarrow \mu^*(A) := \mu^*(\mathbf{1}_A) \in [0, \infty).$$

Obviously, we have

$$\mu^*(\phi) = \mu^*(0) = 0 \text{ and if } A_1 \subset A_2, \text{ then } \mathbf{1}_{A_1} \le \mathbf{1}_{A_2} \text{ and } \mu^*(A_1) \le \mu^*(A_2).$$

For any sequence $(A_n)_n$ of $\mathscr{P}(X)$, we have

$$\mathbf{1}_{\bigcup_{n=1}^{\infty} A_n} \le \sum_{n=1}^{\infty} \mathbf{1}_{A_n}, \ \mu^*\left(\bigcup_{n=1}^{\infty} A_n\right) = \sup_n \mu^*\left(\bigcup_{i=1}^{n} A_i\right) \le \sup_n \sum_{i=1}^{n} \mu^*(A_i),$$

$$\mu^*\left(\bigcup_{n=1}^{\infty} A_n\right) \le \sum_{n=1}^{\infty} \mu^*(A_n).$$

We remember the famous Caratheodory's result:

The family $\mathcal{M}_\mu$ of all subsets $A \subset X$ for which $\mu^*(B) = \mu^*(B \cap A) + \mu^*(B \cap CA)$, $\forall B \in \mathcal{P}(X)$ is a σ-algebra of subsets of X, and the restriction of μ^* at this σ-algebra is countable additive, i.e., it is a positive measure on $\mathcal{M}_\mu$. We shall denote by λ_μ this measure. Having in view that in our case $\mu^*(A) = \mu^*(\mathbf{1}_A) = \inf\{\bar{\mu}(\mathbf{1}_G); \ G - \text{open}, \ A \subset G\}$, $\forall A \in \mathcal{P}(X)$, we deduce that any open subset of X belongs to $\mathcal{M}_\mu$, and therefore any Borel subset of X belongs to $\mathcal{M}_\mu$.

Since any function $\varphi \geq 0$, $\varphi \in C_i$ is the limit of the following increasing sequence $(\varphi_n)_n$ from C_i:

$$\varphi_n = \frac{1}{2^n} \sum_{i-1}^{n \cdot 2^n} \mathbf{1}_{[\varphi > \frac{i}{2^n}]},$$

we deduce that

$$\mu^*(\varphi) = \lim_{n \to \infty} \mu^*(\varphi_n) = \lim_{n \to \infty} \frac{1}{2^n} \sum_{i=1}^{n \cdot 2^n} \mu^* \left(\mathbf{1}_{[\varphi > \frac{i}{2^n}]} \right)$$

$$= \lim_{n \to \infty} \frac{1}{2^n} \sum_{i=1}^{n \cdot 2^n} \lambda_\mu([\varphi > \frac{i}{2^n}]) = \lim_{n \to \infty} \int \varphi_n d\lambda_\mu = \int \varphi d\lambda_\mu.$$

Particularly, for any function $f \in \mathcal{K}(X)$, $f \geq 0$ we have $\mu(f) = \int f d\lambda_\mu$, and therefore for any $f \in \mathcal{K}(X)$ we have $\mu(f) = \int f d\lambda_\mu$. Since any function $\varphi \in C_i$ is of the form $\varphi = f + \varphi'$, $f \in \mathcal{K}(X)$, $\varphi' \in C_i^+$, we get $\bar{\mu}(\varphi) = \int \varphi d\lambda_\mu$. The same equality holds, i.e., $\underline{\mu}(\psi) = \int \psi d\lambda_\mu$, for any $\psi \in C_s$. If we consider a function $f \in \mathcal{L}^1(\mu)$, then there exists an increasing sequence $(\psi_n)_n$ in C_s, and there exists a decreasing sequence $(\varphi_n)_n$ in C_i such that

$$\underline{\mu}(\psi_n) \in \mathbb{R}, \ \bar{\mu}(\varphi_n) \in \mathbb{R}_+, \ \psi_n \leq f \leq \varphi_n, \ \text{and} \ \lim_{n \to \infty} \bar{\mu}(\psi_n) = \mu_*(f) = \mu^*(f) = \inf_n \bar{\mu}(\varphi_n).$$

Hence we have

$$\psi_n \in \mathcal{L}^1(\lambda_\mu), \ \varphi_n \in \mathcal{L}^1(\lambda_\mu) \text{ and } \lim_{n \to \infty} \int \psi_n d\lambda_\mu = \lim_{n \to \infty} \int \varphi_n d\lambda_\mu, \ \lim_{n \to \infty} \int (\varphi_n - \psi_n) d\lambda_\mu = 0.$$

On the other hand, for $n < m$ we have

$$0 \leq \psi_m - \psi_n < \varphi_n - \psi_n, \ \ 0 \leq \varphi_n - \varphi_m \leq \varphi_n - \psi_n,$$

$$\int |\psi_m - \psi_n| d\lambda_\mu = \int (\psi_m - \psi_n) d\lambda_\mu \leq \int (\varphi_m - \psi_n) d\lambda_\mu,$$

$$\int |\varphi_m - \varphi_n| d\lambda_\mu = \int (\varphi_n - \varphi_m) d\lambda_\mu \leq \int (\varphi_n - \psi_n) d\lambda_\mu,$$

and therefore the sequences $(\varphi_n)_n$, $(\psi_n)_n$ are convergent in $\mathcal{L}^1(\lambda_\mu)$ to $\overline{f}$ and to $\underline{f}$, respectively, $\underline{f} \leq \overline{f}$. Since $\int \overline{f} d\lambda_\mu - \int \underline{f} d\lambda_\mu \leq \int (\varphi_n - \psi_n) \, d\lambda_\mu$ for any $n \in \mathbb{N}$ we get that $\underline{f} = \overline{f}$, λ_μ-a.e. Since the starting function f satisfies the inequalities $\psi_n \leq f \leq \varphi_n$, for any $n \in \mathbb{N}$, we get $f = \underline{f} = \overline{f}$, λ_μ -a.e., and therefore $f \in \mathcal{L}^1(\lambda_\mu)$. Moreover,

$$\mu(f) = \lim_{n \to \infty} \underline{\mu}(\psi_n) = \lim_{n \to \infty} \int \psi_n d\lambda_\mu = \int f d\lambda_\mu.$$

In fact, $\mathcal{L}^1(\mu) = \mathcal{L}^1(\lambda_\mu)$ and $\mu(f) = \int f d\lambda_\mu$ for any $f \in \mathcal{L}^1(\mu)$. The only relation to be justified, $\mathcal{L}^1(\lambda_\mu) \subset \mathcal{L}^1(\mu)$, follows from the definition of outer measure $\mu^* : \mathscr{P}(X) \to [0, \infty]$, and from the fact that for any element $A \in \mathcal{M}_\mu$ with $\mu^*(A) < \infty$, we have $\mathbf{1}_A \in \mathcal{L}^1(\mu)$. The last relation is obviously true if A is an open set of X since in this case $\mathbf{1}_A \in C_i$, $\overline{\mu}(\mathbf{1}_A) = \mu^*(\mathbf{1}_A) = \mu^*(A) < \infty$.

Moreover, for any open subset $G \subset X$, we have

$$\mu^*(G) = \overline{\mu}(\mathbf{1}_G) = \sup \left\{ \underline{\mu}(K); \ K \subset G, K - compact \right\}$$

$$= \sup \left\{ \mu^*(K); \ K \subset G, K - compact \right\} = \sup \left\{ \lambda_\mu(K); \ K \subset G, K - compact \right\}.$$

Also, for any compact set $K \subset X$, we have

$$\mathbf{1}_K \in C_s, \ \underline{\mu}(\mathbf{1}_K) < \infty, \ \mathbf{1}_K \in \mathcal{L}^1(\mu) \text{ and}$$

$$\mu^*(K) = \underline{\mu}(\mathbf{1}_K) = \mu_*(\mathbf{1}_K) = \mu^*(\mathbf{1}_K) = \inf \{\overline{\mu}(\mathbf{1}_G); \ K \subset G, \ G - open\}.$$

We show now that for any $A \in \mathcal{M}_\mu$ with $\lambda_\mu(A) < \infty$ we have

$$\lambda_\mu(A) = \sup \left\{ \lambda_\mu(K); \ K \subset A, \ K - compact \right\}.$$

Let G_0 an open subset, $A \subset G_0$, $\lambda_\mu(G_0) < \infty$, and let $(K_n)_n$ an increasing sequence of compact sets, $K_n \subset G_0$ such that $\sup\{\lambda_\mu(K_n); n \in \mathbb{N}^*\} = \lambda_\mu(G_0)$. We also consider a decreasing sequence $(G_n)_n$ of open sets such that

$$G_n \supset G_0 \backslash A, \ \lambda_\mu(G_0 \backslash A) = \inf\{\lambda_\mu(G_n); n \in \mathbb{N}^*\}.$$

We have

$$K_m \backslash G_m \subset G_0 \backslash (G_0 \backslash A) = A, \ \forall m \in \mathbb{N},$$

$$\lambda_\mu(K_m) \leq \lambda_\mu(K_m \backslash G_m) + \lambda_\mu(G_m), \ \forall m \in \mathbb{N},$$

$$\lambda_\mu(K_m) \leq \sup \left\{ \lambda_\mu(K); \ K \subset A, \ K - compact \right\} + \lambda_\mu(G_m),$$

$$\lambda_\mu(K_m) \le \mu_*\,(\mathbf{1}_A) + \lambda_\mu(G_m),$$

$$\lambda_\mu(G_0) \le \mu_*\,(\mathbf{1}_A) + \lambda_\mu(G_0 \backslash A),$$

$$\lambda_\mu(G_0) = \lambda_\mu(A) + \lambda_\mu(G_0 \backslash A),$$

and therefore $\mu_*\,(\mathbf{1}_A) \ge \lambda_\mu A) = \mu^*\,(\mathbf{1}_A)$, i.e., $\mathbf{1}_A \in \mathcal{L}^1(\mu)$. $\square$

Remark 2.2.3 For any Radon measure μ on X there exists a measure $\lambda_\mu : \mathscr{B}(X) \to [0, \infty]$ such that for any Borel subset A of X we have

$$\lambda_\mu(A) = \sup\{\lambda_\mu(K); \;\; K \subset A, \;\; K - \text{compact}\} = \inf\{\lambda_\mu(G); \;\; A \subset G, \;\; G - open\} \text{ and}$$

$$\mu(f) = \int f d\lambda_\mu, \;\; \forall f \in \mathcal{L}^1(\mu) = \mathcal{L}^1(\lambda_\mu).$$

Lemma 2.2.4 *Let v be an upper semicontinuous non-negative real function on X, and let*

$$B_v = \{f \in \mathcal{K}(X); \;\; p_v(f) \le 1\} = \left\{f \in \mathcal{K}(X); \;\; \sup_{x \in X} |f(x)|\, v(x) \le 1\right\},$$

the unit ball of $\mathcal{K}(X)$ associated to the seminorm p_v. The dual of the locally convex space $(\mathcal{K}(X), \; p_v)$ is the linear subspace M_v of all Radon measures μ on $\mathcal{K}(X)$ such that $|\mu|\left(\frac{1}{v}\right) < \infty$. The polar set of B_v with respect to the duality $(\mathcal{K}(X), \; M_v)$, i.e., the set $B_v{}^0 = \{\mu \in M_v; \; \|\mu\| \le 1\}$ is the set of all Radon measures μ on $\mathcal{K}(X)$ with the property $|\mu|\left(\frac{1}{v}\right) \le 1$. This set is compact if we endow the set of all Radon measures $\mathcal{M}(X)$ on X with the weak topology, i.e., the smallest topology on $\mathcal{M}(X)$ making continuous the linear maps on $\mathcal{M}(X)$:

$$\mu \to \mu(f), \;\; \forall f \in \mathcal{K}(X).$$

Proof Indeed, if we consider a linear map $\theta : \mathcal{K}(X) \to \mathrm{K}$, which is continuous with respect to the seminorm $p_v : \mathcal{K}(X) \to \mathrm{K}$, then we have

$$\|\theta\|_v = \sup\{|\theta(f)|; \;\; f \in \mathcal{K}(X), \;\; p_v(f) \le 1\} < \infty.$$

Since, for any compact set K, the lower semicontinuous function $\frac{1}{v} : X \to (0, \infty]$ has a strictly positive infimum α_K on K, we deduce that for any function $\varphi \in \mathcal{K}(X)$ which vanishes outside K we have

$$p_v(\varphi) = \sup\{|v(x) \cdot \varphi(x)|; \ x \in K\} \leq \frac{1}{\alpha_K} \cdot \|\varphi\|, \ \text{i.e.,}$$

$$p_v\left(\frac{\alpha_K}{\|\varphi\|} \cdot \varphi\right) \leq 1, \quad \left|\theta\left(\frac{\alpha_K}{\|\varphi\|} \cdot \varphi\right)\right| \leq \|\theta\|_v, \quad |\theta(\varphi)| \leq \frac{1}{\alpha_K} \cdot \|\theta\|_v \cdot \|\varphi\|.$$

Hence θ is a Radon measure on X. From the definition of the positive measure $|\theta|$, we have for any $f \in \mathcal{K}^+(X)$, $f \leq \frac{1}{v}$,

$$|\theta|(f) = \sup\{|\theta(g)|; \ g \in \mathcal{K}(X), \ |g| \leq f\},$$

$$|\theta|\left(\frac{1}{v}\right) = \sup\left\{|\theta|(f); \ f \in \mathcal{K}^+(X), \ f \leq \frac{1}{v}\right\}$$

$$= \sup\left\{|\theta(g)|; \ g \in \mathcal{K}(X), \ |g| \leq \frac{1}{v}\right\}$$

$$= \sup\{|\theta(g)|; \ g \in \mathcal{K}(X), \ p_v(g) \leq 1\} = \|\theta\|_v.$$

The compacity of the set $B_v{}^0$:

$$B_v{}^0 = \left\{\mu \in \mathcal{M}(X); \ |\mu|\left(\frac{1}{v}\right) \leq 1\right\}$$

with respect to the weak topology on $\mathcal{M}(X)$ given by the duality $(f, \mu) \to \mu(f)$ defined on $\mathcal{K}(X) \times \mathcal{M}(X)$ follows now from Alaoglu's theorem applied to the locally convex space $\mathcal{K}(X)$ endowed with the seminorm p_v. $\qquad\square$

Theorem 2.2.5 *Let V be a Nachbin family on X, and let $CV_0(X)$ be the weighted space associated with family V, endowed with weighted topology ω_V. Then, the dual $CV_0(X)^*$ of the locally convex space $(CV_0(X), \omega_V)$ is identical with the dual of the space $\mathcal{K}(X)$ endowed with the induced ω_V—topology. More precisely, $\theta \in CV_0(X)^*$ iff there exists a Radon measure μ on X and $v \in V$ such that*

$$|\mu|\left(\frac{1}{v}\right) < \infty, \quad CV_0(X) \subset \mathcal{L}^1(|\mu|), \quad \text{and} \quad \theta(f) = \int f d\mu, \ \forall f \in CV_0(X).$$

Proof We known (from Proposition 2.1.4) that $\mathcal{K}(X)$ is dense in the locally convex space $(CV_0(X), \omega_V)$. Hence any element $\theta \in CV_0(X)^*$ is totally determined by its restriction to $\mathcal{K}(X)$. On the other hand, there exists a weight $v \in V$ and $\alpha \in \mathbb{R}_+$ such that

$$|\theta(f)| \leq \alpha \cdot p_v(f), \quad \forall f \in CV_0(X).$$

Hence the restriction of θ to $\mathcal{K}(X)$ satisfies the same inequality, and therefore, using Lemma 2.2.4, there exists a Radon measure μ on X such that

$$|\mu|\left(\frac{1}{v}\right) < \infty, \quad \theta(f) = \int f d\mu, \quad \forall f \in \mathcal{K}(X).$$

Since any function $g \in CV_0(X)$ is dominated to infinity by $\frac{1}{v}$, and $\frac{1}{v}$ is strictly positive on X, we deduce that there exists $\beta \in R_+$ such that $|g| \leq \beta \cdot \frac{1}{v}$, on X, and therefore $g \in \mathcal{L}^1(|\mu|)$. Moreover, there exists a sequence $(f_n)_n \in \mathcal{K}(X)$ such that

$$p_v(g - f_n) \leq \frac{1}{n}, \quad \forall n \in \mathbb{N}^*, \quad \lim_{n \to \infty} \theta(f_n) = \theta(g).$$

Hence we have $|g - f_n| \leq \frac{1}{n} \cdot \frac{1}{v}$, $\forall n \in \mathbb{N}^*$, and therefore,

$$\left| \int (g - f_n) \, d\mu \right| \leq \int |g - f_n| \, d|\mu| \leq \frac{1}{n} \cdot \int \frac{1}{v} d |\mu| = \frac{1}{n} \cdot |\mu| \left(\frac{1}{v}\right),$$

$$\int g d\mu = \lim_{n \to \infty} \int f_n d\mu = \lim_{n \to \infty} \theta(f_n) = \theta(g).$$

Let $\mathcal{B}(X)$ be the σ-algebra of all Borel subsets of X and $\lambda : \mathcal{B}(X) \to \overline{\mathbb{R}}_+$ a measure, i.e., a countable additive function, $\lambda(\phi) = 0$ with $\lambda(K) < \infty$ for any compact subset of X, and such that for any $A \in B(X)$ we have the "regularity" property:

$$\lambda(A) = \sup\{\lambda(K); \quad K \subset A, \quad K - \text{compact}\}.$$

(Such type of countable additive maps are those λ_μ associated with positive Radon measures μ on $\mathcal{K}(X)$). It is obvious that all elements $f \in \mathcal{K}(X)$ are integrable with respect to λ, and that the map $\mu : \mathcal{K}(X) \to \mathbb{R}$:

$$f \to \int f d\lambda = \mu(f)$$

is a Radon measure on $\mathcal{K}(X)$.

Since it is not difficult to show that

$$\lambda(K) = \inf\{\lambda(G); \quad K \subset G, \quad G - \text{open}\}$$

for any compact subset K of X, and since for any open set $G \subset X$ such that $K \subset G$ there exists $f \in \mathcal{K}^+(X)$ with $\mathbf{1}_K \leq f \leq \mathbf{1}_G$, we deduce that

$$\lambda(K) = \inf\{\mu(f); \quad f \in \mathcal{K}(X), \quad f \geq \mathbf{1}_K\} = \underline{\mu}(\mathbf{1}_K) = \lambda_\mu(K).$$

where λ_μ is the countable additive map associated to the Radon measure μ. Hence we have $\lambda(A) = \lambda_\mu(A)$ for any Borel set A, and therefore $\lambda = \lambda_\mu$ on $\mathscr{B}(X)$.

In the sequel, we shall use equally the term "Radon measure" for positive linear functional on the linear space $\mathcal{K}(X)$ and for countable additive regular map on $\mathscr{B}(X)$. The convention will be extended also for signed Radon measures on $\mathcal{K}(X)$ or countable additive maps $\lambda : \mathscr{B}(X) \to \mathbb{K}$.

If μ is a Radon measure, and $f : X \to \overline{\mathbb{R}}$ is $|\mu|$-integrable, then we denote by $f \cdot \mu$ the Radon measure on X given by

$$(f \cdot \mu)(A) = \int_A f d\mu = \int f \cdot \mathbf{1}_A d\mu, \quad \forall A \in \mathscr{B}(X), \text{ or}$$

$$(f \cdot \mu)(g) = \int f \cdot g d\mu, \quad \forall f \in \mathcal{K}(X).$$

We have $|f \cdot \mu| = |f| \cdot |\mu|$, and therefore $|f \cdot \mu|(\mathbf{1}_K) \leq \int |f| d |\mu| < \infty$, i.e., the Radon measure $f \cdot \mu$ on X is bounded:

$$|(f \cdot \mu)(g)| \leq \alpha \cdot \|g\|, \quad \forall g \in \mathcal{K}(X), \text{ where } \alpha = \int |f| d |\mu|.$$

$\square$

Having in view these conventions, the previous theorem may be expressed as follows.

Theorem 2.2.6 *The dual of the locally convex space $(CV_0(X), \omega_V)$ is the set:*

$$V \cdot \mathcal{M}_b(X) = \{v \cdot \lambda; \ v \in V, \ \lambda \in \mathcal{M}_b(X)\},$$

where $M_b(X)$ is the set of all bounded Radon measures on X.
    More exactly, for any $v \in V$ we have

$$B_v{}^0 = \left\{\mu \in \mathcal{M}(X); \ |\mu|\left(\frac{1}{v}\right) \leq 1\right\} = \{v \cdot \lambda; \ \lambda \in M_b(X), \ \|\lambda\| \leq 1\}.$$

If $\mu \in \mathcal{M}^+(X)$, then there exists a smallest closed subset $F \subset X$ such that $\mu(X \backslash F) = 0$. Indeed, the family $\mathcal{O}_\mu = \{G \subset X; \ \mu(G) = 0, \ G - \text{open}\}$ is upper directed with respect to the inclusion order relation, and the set $G_0 = \bigcup_{G \in \mathcal{O}_\mu} G$ belongs to $\mathcal{O}_\mu$ because

$$\mu(G_0) = \sup \{\mu(K); \ K \subset G_0, \ K - \text{compact}\} = 0.$$

The last equality comes from the fact that for any compact K, $K \subset G_0$, there exists $G \in \mathcal{O}_\mu$ such that $K \subset G$. Obviously, the closed subset $\mathrm{supp}\mu = X \backslash G_0$ is the smallest closed subset F of X for which $\mu(X \backslash F) = 0$. This set is called the support of the measure μ. Clearly, we have

$$\mu = \mathbf{1}_{\mathrm{supp}_\mu} \cdot \mu.$$

For an arbitrary Radon measure $\mu \in \mathcal{M}(X)$, we put

$$\mathrm{supp}\mu = \mathrm{supp}\, |\mu|,$$

and we have

$$\mu = \mathbf{1}_{\mathrm{supp}\mu} \cdot \mu.$$

Theorem 2.2.7 *The dual of the space $C_b(X)$ (of all scalar continuous and bounded functions on X) endowed with the strict topology β is just the space $\mathcal{M}_b(X)$ of all bounded Radon measures on X.*

Proof We recall that the strict topology β is the weighted topology ω_V corresponding to the Nachbin family $V = C_0^+(X)$—the nonnegative continuous functions on X which vanish at infinity and $CV_0(X) = C_b(X)$ (see Example 2.1.8).

By Theorem 2.2.6, the dual of the locally compact space $(C_b(X), \beta)$ is

$$V \cdot \mathcal{M}_b(X) = C_0^+(X) \cdot \mathcal{M}_b(X).$$

Obviously, $C_0^+(X) \cdot \mathcal{M}_b(X) \subset \mathcal{M}_b(X)$. The converse inclusion is also true. Let us take a positive, bounded Radon measure μ on X. Since $\mu(X) < \infty$ and $\mu(X) = \sup\{\mu(K); K \subset X, K - \text{compact}\}$, we may consider an increasing sequence $(K_n)_n$ of compact subsets of X such that $K_n \subset \overset{\circ}{K}_{n+1}$, $\forall n \in \mathbb{N}$, and such that

$$\sup_n \mu(K_n) = \mu(X) < \infty, \lim_{n \to \infty} \mu(X \backslash K_n) = 0.$$

Passing eventually to a subsequence of sequence $(K_n)_n$, we may suppose that

$$\mu(X \backslash K_n) < \frac{1}{2^n} \, for all \, n \in \mathbb{N}.$$

Since for any $n \in \mathbb{N}$ the closed subset $X \backslash \overset{\circ}{K}_{n+1}$ and the compact subset K_n are disjoint, by Uryson's Theorem, there exists a continuous function $f_n : X \to [0, 1]$ such that its restriction to $X \backslash \overset{\circ}{K}_{n+1}$ is equal to 1 and f_n vanishes on K_n. From the construction, we have

$$\mathbf{1}_{X \backslash \overset{\circ}{K}_{n+1}} \leq f_n \leq \mathbf{1}_{X \backslash K_n},$$

$$\mu(f_n) \leq \frac{1}{2^n}, \ \forall n \in \mathbb{N},$$

$$\mu\left(\sum_{n=1}^{\infty} f_n\right) \leq 1,$$

$$\mu\left(1 + \sum_{n=1}^{\infty} f_n\right) \leq \mu(X) + 1 < \infty.$$

The function $f : X \to [0, \infty]$, given by : $f = 1 + \sum_{n=1}^{\infty} f_n$, is obviously lower semi-continuous as a sum of positive continuous functions, $f(x) = \infty$ for any $x \in X \backslash \bigcup_n (\overset{\circ}{K}_{n+1}) = X \backslash \bigcup_n K_n$, and therefore f is continuous at any $x \in X \backslash \bigcup_n K_n$. Since $f_n \leq \mathbf{1}_{X \backslash K_n}$, $\forall n \in \mathbb{N}$, we deduce that $f_n = 0$ on K_n, and therefore $f_m = 0$ on K_n for any $m \geq n$. Hence we have

$$f(x) = 1 + \sum_{i<n} f_i(x), \ \ \forall x \in K_n, \ \ f(x) = 1 + \sum_{i \leq n} f_i(x), \ \ \forall x \in \overset{\circ}{K}_n,$$

f is continuous on $\overset{\circ}{K}_n$, i.e., f is continuous on $\bigcup \overset{\circ}{K}_n = \bigcup K_n$, so f is continuous on X.
From the inequality $\mathbf{1}_{X \backslash \overset{\circ}{K}_{n+1}} \leq f_n$ for any $n \in \mathbb{N}$ we get

$$x \notin K_m \ \Rightarrow \ x \notin K_n, \ \forall n \leq m \ \Rightarrow \ x \notin \overset{\circ}{K}_n, \ \forall n \leq m,$$

$$\sum_{n \leq m} f_n(x) \geq \sum_{n+1 \leq m} \mathbf{1}_{X \backslash \overset{\circ}{K}_{n+1}} \geq m - 1, \ \ x \notin K_m \ \Rightarrow \ f(x) \geq m.$$

Hence the function $g(x) = \frac{1}{f(x)}$ belongs to $C_0(X)$, $f \cdot \mu \in \mathcal{M}_b(X)$, $\mu = g \cdot (f \cdot \mu) \in V \cdot \mathcal{M}_b(X)$. $\qquad \square$

2.3   Stone–Weierstrass Theorem in Weighted Spaces

Let X be a Hausdorff locally compact space, and let $\mathcal{F}$ be a subset of continuous functions on X with values in the interval $[0, 1]$. The set $\mathcal{F}$ induces on X the following equivalence relation:

$$x \sim_{\mathcal{F}} y \;\Leftrightarrow\; f(x) = f(y), \forall f \in \mathcal{F}.$$

For any $x \in X$ we denote by $[x]_{\mathcal{F}}$ the set:

$$[x]_{\mathcal{F}} = \{y \in X;\ f(y) = f(x), \forall f \in \mathcal{F}\}.$$

Obviously, $[x]_{\mathcal{F}}$ is a closed subset of X and any element $f \in \mathcal{F}$ is constant on this set. In fact $[x]_{\mathcal{F}}$ is the maximal set containing x with this property, and for any $x, y \in X$ we have $[x]_{\mathcal{F}} = [y]_{\mathcal{F}}$ or $[x]_{\mathcal{F}} \cap [y]_{\mathcal{F}} = \phi$. We shall denote by $X/\sim_{\mathcal{F}}$ the space of all equivalence classes, i.e.,

$$X/\sim_{\mathcal{F}} = \{[x]_{\mathcal{F}};\ x \in X\}.$$

Furthermore, we denote by βX the Stone-Čech compactification of the locally compact space X, i.e., the compact space (uniquely determined up to a topological isomorphism) such that X is a dense subset of βX. The topology of X coincides with the trace topology of βX on X, and any bounded continuous function on X may be extended to a unique continuous function on βX.

If for any $f \in C_b(X)$ (i.e., f is a continuous bounded function on X) we shall denote by βf the continuous extension of f to βX, then the map $\beta : C_b(X) \to C(\beta X)$ is an isomorphism between two Banach algebras.

The family $\beta \mathcal{F} = \{\beta f;\ f \in \mathcal{F}\}$ induces on βX the following equivalence relation:

$$u \sim_{\beta \mathcal{F}} v \;\Leftrightarrow\; (\beta f)(u) = (\beta f)(v), \forall f \in \mathcal{F}.$$

We denote by Z the quotient space $\beta X/\sim_{\beta \mathcal{F}}$ and by $\pi : \beta X \to Z$ the canonical mapping.

For any $f \in \mathcal{F}$ the function $\tilde{f} : Z \to \mathrm{K}$ is given by $\tilde{f}(z) = (\beta f)(y)$ if $z = \pi(y)$. Clearly, the function $\tilde{f}$ is well-defined. If we denote by $\tilde{\mathcal{F}} = \left\{\tilde{f}\right\}_{f \in \mathcal{F}}$, then $\tilde{\mathcal{F}} \subset C(Z)$.

Theorem 2.3.1 *Let* $\mathcal{F} \subset C(X, [0, 1])$ *with the property that* $\tilde{\mathcal{F}} = C(Z, [0, 1])$. *We suppose in addition that for any* $x \in X$ *there exists a compact subset of* $K_x \subset X$ *such that* $[x]_{\mathcal{F}} \cap K_x = \phi$. *Then, there exists a finite number of equivalence classes* $[x_1]_{\mathcal{F}}, [x_2]_{\mathcal{F}}, \ldots, [x_n]_{\mathcal{F}}$ *and there exists a finite number of functions* $f_1, f_2, \ldots, f_n \in \mathcal{F}$

with the properties:

$$f_i \left| K_{x_i} = 0, i = \overline{1, n}, \quad \sum_{i=1}^{n} f_i = 1.\right.$$

Proof We remark that for any $x \in X$ we have $[x]_{\mathcal{F}} = \pi(x) \cap X$, and if $y \in \beta X$ is such that $\pi(y) \cap X \neq \phi$, then we have $\pi(y) \cap X = [x]_{\mathcal{F}}$ for any $x \in \pi(y) \cap X$. Hence π may be seen as an injection from $\{[x]_{\mathcal{F}}; x \in X\}$ into Z, namely

$$\pi\left([x]_{\mathcal{F}}\right) = \pi(y), \quad \forall y \in [x]_{\mathcal{F}}.$$

In fact, we have a bijection between the space $\{[x]_{\mathcal{F}}; x \in X\}$ and the subspace $\pi(X)$ of Z. Since the set $[x]_{\mathcal{F}} = \pi(y) \cap X$ and the compact subset K_x are disjoint, it follows that $\pi(x)$ does not belong to the compact subset $\pi(K_x)$ of $\pi(X)$. This implies that $\bigcap_{x \in X} \pi(K_x) = \phi$. Since $\pi(K_x)$ is compact, we deduce that there exists a finite subset $\{x_1, \ldots, x_n\} \subset X$ such that $\bigcap_{i=1}^{n} \pi(K_{x_i}) = \phi$, and therefore the compact space Z is covered by a finite family of open subsets of Z:

$$Z = \bigcup_{i=1}^{n} \left(Z \backslash \pi(K_{x_i})\right).$$

We consider now a partition $\sum_{i=1}^{n} g_i$ of the constant function $\mathbf{1}$, $g_i : Z \to [0, 1]$, g_i vanishing outside $Z \backslash \pi(K_{x_i})$, g_i continuous for any $i = \overline{1, n}$.

Since $\tilde{\mathcal{F}} = C(Z, [0, 1])$, it follows that there exists $f_i \in \mathcal{F}$ such that $g_i = \tilde{f}_i$, $\forall i = \overline{1, n}$. Clearly, $\tilde{f}_i \circ \pi = \beta f_i$ and that for any $x \in X$ we have

$$f_i(x) = (\beta f_i)(x) = (\tilde{f}_i \circ \pi)(x) = g_i[\pi(x)] \geq 0;$$

$$f_i \left| K_{x_i} = g_i \left| \left(\pi[K_{x_i}]\right) = 0 \text{ and } \sum_{i=1}^{n} f_i(x) = \sum_{i=1}^{n} g_i[\pi(x)] = 1.\right.\right.$$

$\square$

Corollary 2.3.2 (See Nachbin [19, 20], Lemma 1 in [1]) *Let X be a Hausdorff locally compact space, and let $\mathcal{A} \subset C_b(X)$ be a closed subalgebra which contains the constant functions (self-adjoint in the complex case). We suppose in addition that for any $x \in X$ there exists a compact subset $K_x \subset X$ such that $[x]_{\mathcal{A}} \cap K_x = \phi$. Then, there exists a finite number set of equivalence classes $[x_1]_{\mathcal{A}}, [x_2]_{\mathcal{A}}, \ldots, [x_n]_{\mathcal{A}}$ and also there exists a finite* *set of functions* $\{a_1, a_2, \ldots, a_n\} \subset \mathcal{A}$ *with the properties:*

$$a_i \geq 0, \quad a_i \big| K_{x_i} = 0, i = \overline{1, n}, \quad \sum_{i=1}^{n} a_i = 1.$$

Proof First, we remark that if we denote by $\mathcal{A}_1 = \{a \in \mathcal{A}; \ 0 \leq a \leq 1\}$, then $x {\sim}_{\mathcal{A}} y$ iff $x {\sim}_{\mathcal{A}_1} y$, i.e., $a(x) = a(y), \forall a \in \mathcal{A}$ iff $b(x) = b(y), \ \forall b \in \mathcal{A}_1$. It is sufficient we show that $b(x) = b(y), \ \forall b \in \mathcal{A}_1$ involving $a(x) = a(y), \forall a \in \mathcal{A}$. Indeed, if $a \in \mathcal{A}_+$, then $b = \frac{a}{\|a\|} \in \mathcal{A}_1$ and so $a(x) = a(y)$. On the other hand, we remark that $\mathcal{A}$ is a lattice since it is a closed subalgebra containing the constant functions.

Therefore for any $a \in \mathcal{A}$, we have $a_+ = \frac{|a|+a}{2} \in \mathcal{A}_+$, $a_- = \frac{|a|-a}{2} \in \mathcal{A}_+$, and hence

$$a(x) = a_+(x) - a_-(x) = a_+(y) - a_-(y) = a(y).$$

If $\mathcal{A} \subset C_b(X, \mathbb{C})$ is self-adjoint then from Remark 1.3.5 it results that $Re(\mathcal{A}) \subset \mathcal{A}$, $Re(\mathcal{A})$ is an algebra, and $\mathcal{A} = Re(\mathcal{A}) + i \cdot Re(\mathcal{A})$. On the other hand, since $\mathcal{A}$ is closed, it follows that $Re(\mathcal{A})$ is also a closed algebra. Hence $a(x) = a(y), \ \forall a \in Re(\mathcal{A})$ and so $a(x) = a(y), \ \forall a \in \mathcal{A}$.

Now, the proof of the Corollary 2.3.2 follows from Theorem 2.3.1 for $\mathcal{F} = \mathcal{A}_1 = \{a \in \mathcal{A}; \ 0 \leq a \leq 1\}$.

Indeed, $\tilde{\mathcal{A}}$ is a closed subalgebra which contains the constant functions and separates the points of the compact set Z. Therefore, by Stone–Weierstrass theorem, we have $\tilde{\mathcal{A}} = C(Z)$. Obviously, $\tilde{\mathcal{A}}_1 = C(Z, [0, 1])$.

We recall that a subset $\mathcal{M} \subset C(X, [0, 1])$ has the property VN if

$$f \cdot g + (1 - f) \cdot h \in \mathcal{M}, \forall f, g, h \in \mathcal{M}. \qquad \square$$

Corollary 2.3.3 *Let X be a Hausdorff locally compact space and let $\mathcal{M} \subset C(X, [0, 1])$ be a closed subset with the property VN which contains the constant functions 0, 1 and at least a constant function $0 < c < 1$. We suppose, in addition, that for any $x \in X$ there exists a compact subset $K_x \subset X$ such that $[x]_{\mathcal{M}} \cap K_x = \phi$. Then, there exists a finite number of equivalence classes $[x_1]_{\mathcal{M}}, [x_2]_{\mathcal{M}}, \ldots, [x_n]_{\mathcal{M}}$ and there exists a finite set of functions $m_1, m_2, \ldots, m_n \in \mathcal{M}$ with the properties:*

$$m_i \big| K_{x_i} = 0, i = \overline{1, n}, \quad \sum_{i=1}^{n} m_i = 1.$$

Proof The proof results from Theorem 2.3.1 for $\mathcal{F} = \mathcal{M}$. Indeed, the set $\tilde{\mathcal{M}}$ is a closed subset of $C(Z, [0, 1])$ which separates the points of the compact set $Z = \beta X / {\sim}_{\beta \mathcal{M}}$ and contains the constant functions 0, 1 and at least a constant function $0 < c < 1$. From Theorem 4.18 in [2], it follows that $\tilde{\mathcal{M}} = C(Z, [0, 1])$. $\qquad \square$

Corollary 2.3.4 *Let X be a Hausdorff locally compact space, and let $C \subset C_b{}^+(X)$ be a closed convex cone containing the constant functions 0, 1 and having the property that for any $u, v \in \beta X$, $\pi(u) \neq \pi(u)$ there is some $\varphi \in C(X, [0, 1])$ such that $\varphi \cdot f + (1-\varphi) \cdot h \in C$, $\forall f, h \in C$ and $(\beta\varphi)(u) \neq (\beta\varphi)(v)$. We suppose also that for any $x \in X$ there exists a compact subset $K_x \subset X$ such that $[x]_C \cap K_x = \phi$.*

Then, there exists a finite number of equivalence classes $[x_1]_C$, $[x_2]_C$, ..., $[x_n]_C$ and an equal number of functions $h_1, h_2, \ldots, h_n \in C$ with the properties:

$$h_i \big| K_{x_i} = 0, i = \overline{1, n}, \quad \sum_{i=1}^{n} h_i = 1.$$

Proof The proof results from Theorem 2.3.1 for $\mathcal{F} = \mathcal{C}_1 = \{h \in \mathcal{C}; \ h \leq 1\}$. Indeed, $\tilde{\mathcal{C}}$ is a closed convex cone of $C^+(Z)$ which contains the constant functions 0, 1, separates the points of the compact set Z, and has the property that for any $z_1, z_2 \in Z$, $z_1 \neq z_2$, there is a multiplier $\tilde{\varphi} \in C(Z, [0, 1])$, (i.e., $\tilde{\varphi} \cdot \tilde{f} + (1 - \tilde{\varphi}) \cdot \tilde{h} \in \tilde{\mathcal{C}}, \forall \tilde{f}, \tilde{h} \in \tilde{\mathcal{C}}$) with the property $\tilde{\varphi}(z_1) = (\beta\varphi)(u) \neq (\beta\varphi)(v) = \tilde{\varphi}(z_2)$ where $z_1 = \pi(u)$, $z_2 = \pi(v)$. From Boboc and Bucur (see Theorem 2, Corollary 1 in [3]), it follows that $\tilde{\mathcal{C}} = C^+(Z)$, and hence that $\tilde{\mathcal{C}}_1 = C(Z, [0, 1])$.

Furthermore, we consider a Nachbin family V on X and we denote by $CV_0(X, \mathbb{R})$ or, simpler, by $CV_0(X)$ the weighted space corresponding to the family V. We recall that any weight $v \in V$ generates a seminorm $p_v : CV_0(X) \to \mathbb{R}_+$ defined by

$$p_v(f) = \sup\{v(x) \cdot |f(x)| ; \ x \in X\}, \quad \forall f \in CV_0(X).$$

The locally convex topology defined by this family of seminorms is denoted by ω_V and it is called the weighted topology on $CV_0(X)$.

If for any $x \in X$ there exists $v_x \in V$ such that $v_x(x) > 0$, then $(CV_0(X), \omega_V)$ is a Hausdorff locally convex space. Furthermore, we suppose that $CV_0(X)$ is Hausdorff. $\square$

Definition 2.3.5 A linear subspace $\mathcal{W}$ of $CV_0(X)$ is called localizable with respect to the family $\mathcal{F}$ of $C_b(X)$ if

$$\overline{\mathcal{W}} = \left\{ f \in CV_0(X); \ f \big| [x]_{\mathcal{F}} \in \overline{\mathcal{W}|[x]}_{\mathcal{F}}, \forall x \in X \right\}.$$

Remark 2.3.6 The linear subspace $\mathcal{W} \subset CV_0(X)$ is dense in $CV_0(X)$ if the following conditions are satisfied:

The linear subspace $\mathcal{W} \subset CV_0(X)$ is dense in $CV_0(X)$ if the following conditions are satisfied:

(a) $\mathcal{W}$ is localizable with respect to $\mathcal{F}$,
(b) $\mathcal{F}$ separates the points of X,
(c) For any $x \in X$ there exists a $w \in \mathcal{W}$ such that $w(x) \neq 0$.

Indeed, from (b) it follows that $[x]_{\mathcal{F}} = \{x\}, \forall x \in X$. Let $f \in CV_0(X)$ be arbitrary. If $f(x) = 0$, then obviously $f\left|[x]_{\mathcal{F}} \in \overline{W}\left|[x]_{\mathcal{F}}\right.\right.$. If $f(x) \neq 0$, then by (c) there exists $w \in W$ such that $w(x) \neq 0$.

Furthermore, we have

$$f\left|[x]_{\mathcal{F}} = f(x) = \lambda \cdot w(x) = \lambda \cdot w\left|[x]_{\mathcal{F}} \in \overline{W}\left|[x]_{\mathcal{F}}\right.\right.\right., \text{ where } \lambda = \frac{f(x)}{w(x)}.$$

Therefore, from (b) and (c) we deduce

$$CV_0(X) = \left\{ f \in CV_0(X); \ f\left|[x]_{\mathcal{F}} \in \overline{W}\left|[x]_{\mathcal{F}}, \forall x \in X\right.\right.\right\}.$$

Finally, from (a) we have

$$\overline{W} = CV_0(X).$$

Theorem 2.3.7 *Let $\mathcal{F} \subset C(X; [0, 1])$ be a subset with the property $\tilde{\mathcal{B}} = C(Z, [0, 1])$, where $\mathcal{B} = \overline{\mathcal{F}}$ is the closure of $\mathcal{F}$ in $C(X; [0, 1])$. If $W \subset CV_0(X)$ is a linear subspace with the property $\mathcal{F} \cdot W \subset W$, then W is localizable with respect to M, i.e.,*

$$\overline{W} = \left\{ f \in CV_0(X); \ f\left|[x]_{\mathcal{F}} \in \overline{W}\left|[x]_{\mathcal{F}}, \forall x \in X\right.\right.\right\}.$$

Proof Since it is obvious that the set of the left side of the equality belongs to the set of the right side, it is sufficient to prove the inverse inclusion.

Let $g \in CV_0(X)$ such that $g \mid [x]_{\mathcal{F}} \in \overline{W}\left|[x]_{\mathcal{F}}, \forall x \in X\right.$. We shall prove that $g \in \overline{W}$.

Let $v \in V$ and $\varepsilon > 0$ be arbitrarily fixed. Then, for any $x \in X$ there exists $w_x \in W$ such that

$$v(y)\left|g(y) - w_x(y)\right| < \varepsilon, \forall y \in [x]_{\mathcal{F}}.$$

If we denote by $K_x = \{y \in X; \ v(y)\left|g(y) - w_x(y)\right| \geq \varepsilon\}$, then K_x is a compact set and

$$[x]_{\mathcal{F}} \cap K_x = \phi, [x]_{\mathcal{F}} \cup K_x = X.$$

Since by hypothesis $\tilde{\mathcal{B}} = C(Z, [0, 1])$ where $\mathcal{B} = \overline{\mathcal{F}}$, from Theorem 2.3.1, it follows that there exists a finite number $[x_1]_{\mathcal{F}}, [x_2]_{\mathcal{F}}, \ldots, [x_n]_{\mathcal{F}}$ of equivalence classes and there also exists a finite number of functions $b_1, b_2, \ldots, b_n \in \mathcal{B}$ with the properties:

$$b_i \left| K_{x_i} = 0, \ i = \overline{1, n}, \ \sum_{i=1}^{n} b_i = 1.\right.$$

Furthermore, for any $i \in \{1, 2, \ldots, n\}$ we have

$$b_i(y) \cdot v(y) \cdot \left| g(y) - w_{x_i}(y) \right| \le \varepsilon \cdot b_i(y), \forall y \in X. \tag{2.1}$$

Indeed, if $y \in [x_i]_{\mathcal{F}}$ then we have $v(y) \cdot \left| g(y) - w_{x_i}(y) \right| < \varepsilon$, and if $x \notin [x_i]_{\mathcal{F}}$, then $b_i(y) = 0$.

From (1.1) it follows

$$v(y) \cdot \sum_{i=1}^{n} b_i(y) \cdot \left| g(y) - w_{x_i}(y) \right| < \varepsilon \cdot \sum_{i=1}^{n} b_i(y) = \varepsilon, \forall y \in X.$$

Furthermore, we have

$$v(y) \cdot \left| g(y) - \sum_{i=1}^{n} b_i(y) \cdot w_{x_i}(y) \right|$$

$$= v(y) \cdot \left| \sum_{i=1}^{n} b_i(y) \cdot [g(y) - w_{x_i}(y)] \right| \le v(y)$$

$$\cdot \sum_{i=1}^{n} b_i(y) \cdot \left| g(y) - w_{x_i}(y) \right| \le \varepsilon, \forall x \in X$$

Since $b_1, b_2, \ldots, b_n \in \overline{\mathcal{F}}$, it results that for any $i \in \{1, \ldots, n\}$ and any $\delta > 0$ there is $f_i \in \mathcal{F}$, such that

$$|b_i(y) - f_i(y)| < \delta, \forall y \in X.$$

Since the functions $v \cdot w_{x_i}$ vanish at infinity, it follows that these are bounded on X, and therefore there exists

$$\alpha_i = \sup \left\{ v(y) \cdot \left| w_{x_i} \right| ; y \in X \right\}, i \in \overline{1, n}.$$

Furthermore, we have

$$v(y) \cdot \left| \sum_{i=1}^{n} f_i(y) \cdot w_{x_i}(y) - g(y) \right| \le v(y) \cdot \left| \sum_{i=1}^{n} f_i(y) \cdot w_{x_i}(y) - \sum_{i=1}^{n} b_i(y) \cdot w_{x_i}(y) \right|$$

$$+ v(y) \cdot \left| \sum_{i=1}^{n} b_i(y) \cdot w_{x_i}(y) - g(y) \right| \le \sum_{i=1}^{n} |f_i(y) - b_i(y)| \cdot v(y) \cdot \left| w_{x_i}(y) \right| + \varepsilon \le \delta \cdot \sum_{i=1}^{n} \alpha_i + \varepsilon.$$

If we suppose that

$$\delta < \frac{\varepsilon}{\sum_{i=1}^{n} \alpha_i},$$

then

$$v(y) \cdot \left| \sum_{i=1}^{n} f_i(y) \cdot w_{x_i}(y) - g(y) \right| \le 2 \cdot \varepsilon, \forall y \in X.$$

Finally, if we denote by $w = \sum_{i=1}^{n} f_i \cdot w_{x_i}$, then $w \in \mathcal{F} \cdot \mathcal{W} \subset \mathcal{W}$ and so the proof is finished.   □

Corollary 2.3.8 *If we suppose in addition that $\mathcal{F}$ separates the points of X, and that for any $x \in X$ there exists $a, w \in \mathcal{W}$ such that $w(x) \ne 0$, then $\mathcal{W}$ is dense in $CV_0(X)$, i.e.,*

$$\overline{\mathcal{W}} = CV_0(X).$$

The proof follows from Theorem 2.3.7 and from Remark 2.3.6.

Theorem 2.3.9 (Nachbin [19, 20]) *Let $\mathcal{A}$ be a subalgebra of $C_b(X)$ containing the constant function 1, self-adjoint in the complex case, and let $\mathcal{W} \subset CV_0(X)$ be a linear subspace such that $\mathcal{A} \cdot \mathcal{W} \subset \mathcal{W}$.*
  Then, $\mathcal{W}$ is localizable with respect to $\mathcal{A}$, i.e.,

$$\overline{\mathcal{W}} = \left\{ f \in CV_0(X); \ f \left| [x]_{\mathcal{A}} \in \overline{\mathcal{W} | [x]}_{\mathcal{A}}, \forall x \in X \right. \right\}.$$

The proof follows from Theorem 2.3.7 and Corollary 2.3.8 for $\mathcal{F} = \mathcal{A}_1 = \{a \in \mathcal{A} \ ; \ 0 \le a \le 1\}$.

Corollary 2.3.10 *If we suppose in addition that $\mathcal{A}$ separates the points of X, and that for any $x \in X$ there exists $a, w \in \mathcal{W}$ such that $w(x) \ne 0$, then $\mathcal{W}$ is dense in $CV_0(X)$, i.e.,*

$$\overline{\mathcal{W}} = CV_0(X).$$

Theorem 2.3.11 *Let $\mathcal{M} \subset C(X; [0, 1])$ be a subset with VN property which contains the constant functions 0, 1 and at least a constant function $0 < c < 1$. If $\mathcal{W} \subset CV_0(X)$ is a linear subspace with the property $\mathcal{M} \cdot \mathcal{W} \subset \mathcal{W}$, then $\mathcal{W}$ is localizable with respect to $\mathcal{M}$, i.e.,*

$$\overline{\mathcal{W}} = \left\{ f \in CV_0(X); \ f \left| [x]_{\mathcal{M}} \in \overline{\mathcal{W} | [x]}_{\mathcal{M}}, \forall x \in X \right. \right\}.$$

The proof follows from Theorem 2.3.7 and from Corollary 2.3.3 for $\mathcal{F} = \mathcal{M}$.

Corollary 2.3.12 *If we suppose in addition that $\mathcal{M}$ separates the points of X, and that for any $x \in X$ there exists $w \in \mathcal{W}$ such that $w(x) \neq 0$, then $\mathcal{W}$ is dense in $CV_0(X)$, i.e.,*

$$\overline{\mathcal{W}} = CV_0(X).$$

Theorem 2.3.13 *Let X be a Hausdorff locally compact space and let $\mathcal{C} \subset C_b{}^+(X)$ be a convex cone containing the constant functions 0, 1 and havings the property that for any $u, v \in \beta X$ such that $\pi(u) \neq \pi(v)$, there is a multiplier $\varphi \in C(X, [0, 1])$ (i.e., $\varphi \cdot f + (1 - \varphi) \cdot h \in \mathcal{C}, \forall f, h \in \mathcal{C}$) such that $(\beta\varphi)(u) \neq (\beta\varphi)(v)$, and let $\mathcal{W} \subset CV_0(X)$ be a linear subspace such that $\mathcal{C} \cdot \mathcal{W} \subset \mathcal{W}$. Then, $\mathcal{W}$ is localizable with respect to $\mathcal{C}$, i.e.,*

$$\overline{\mathcal{W}} = \left\{ f \in CV_0(X);\ f\,|[x]_{\mathcal{C}} \in \overline{\mathcal{W}\,|[x]}_{\mathcal{C}},\ \forall x \in X \right\}.$$

The proof follows from Theorem 2.3.7 and from Corollary 2.3.3 for $\mathcal{F} = \mathcal{C}$.

Corollary 2.3.14 *Let $\mathcal{C} \subset C_b{}^+(X)$ and $\mathcal{W} \subset CV_0(X)$ be as in Theorem 2.3.13. If we suppose in addition that $\mathcal{C}$ separates the points of X, and that for any $x \in X$ there exists $w \in \mathcal{W}$ such that $w(x) \neq 0$, then $\mathcal{W}$ is dense in $CV_0(X)$, i.e.,*

$$\overline{\mathcal{W}} = CV_0(X)$$

The assertion follows from Theorem 2.3.13 and Remark 2.3.6.

Furthermore, X will be a Hausdorff locally compact space, and V will be a Nachbin family of weights on X. For any subset Y of X and any subset W of $CV_0(X)$, we denote by $\overline{W\,|Y}$ the set of all functions $g : Y \to K$ such that for any $v \in V$ there exists a sequence $(f_n)_n$ in W, such that the sequence $(v \cdot f_n)_n$ converges uniformly on Y to the function $v \cdot g$, i.e.,

$$\lim_{n \to \infty} \sup_{x \in Y} |v(x) \cdot [f_n(x) - g(x)]| = 0.$$

It is easy to show that for any scalar function $g : Y \to K$ we have: $g \in \overline{W\,|Y}$ iff there exists a generalized sequence $(f_i)_i$ in W such that $f_i \xrightarrow{\omega_V} g$ on Y.

Lemma 2.3.15 *Let L be a linear functional on $CV_0(X)$, ω_V-continuous, and let $F \subset X$ be a Borel subset such that $|L|\,(X \backslash F) = 0$, where $|L|$ is the modulus of the Radon measure L on X.*

(a) *If $(f_i)_i$ is a generalized sequence in $CV_0(X)$ such that $f_i \xrightarrow{\omega_V} g$ on F, then $g \in \mathcal{L}^1(L)$ and $\lim_i L(f_i) = \int g\, dL$.*

(b) *Let $(f_n)_n$ be a sequence of $CV_0(X)$ which is ω_V-bounded on F, i.e., for any $v \in V$ there exists $\alpha_V \in \mathbf{R}_+$ such that $|v(x) \cdot f_n(x)| \leq \alpha_v,\ \forall x \in F,\ \forall n \in \mathbb{N}$. If the sequence*

$(f_n)_n$ *is pointwise convergent on* F *to a function* g, *then* $g \in \mathcal{L}^1(L)$ *and* $\lim\limits_{n \to \infty} L(f_n) = \int g\,dL$.

Proof   Since $L \in [CV_0(X)]^*$, there exists $v \in V$ such that $|L(f)| \leq p_v(f)$, and therefore by Theorem 2.2.5, L is a Radon measure on X with $|L|\left(\frac{1}{v}\right) < \infty$.

(a) Since $f_i \xrightarrow{\omega_V} g$ on F, *for any* $\varepsilon > 0$, *there exists* $i_\varepsilon \in I$ *such that*

$$v(x) \cdot |f_i(x) - g(x)| < \varepsilon, \ \forall x \in F, \ \forall i \geq i_\varepsilon.$$

Taking $\varepsilon = \frac{1}{n}$, $n \in \mathbb{N}^*$, we may consider an increasing sequence $(i_n)_n$ in I such that

$$v(x) \cdot |f_i(x) - g(x)| \leq \frac{1}{n}, \ \forall x \in F, \ \forall i \geq i_n.$$

Particularly, we have

$$\left|f_{i_n} - g\right| \leq \frac{1}{n} \cdot \frac{1}{v} \text{ on } F,$$

and therefore the sequence $\left(f_{i_n}\right)_n$ is pointwise convergent to g on the set $F \cap [v > 0]$. Since $X \backslash (F \cap [v > 0]) = (X \backslash F) \cup [v = 0]$ and $|L|(X \backslash F) = 0$, $|L|([v = 0]) = 0$, we deduce that the function g belongs to $\mathcal{L}^1(|L|)$, and we have

$$\left|L(f_i) - \int g\,dL\right| = \left|\int (f_i - g)dL\right| \leq \int |f_i - g|\,dL \leq \frac{1}{n} \cdot \int \frac{1}{v}d\,|L|,$$

for any $i \geq i_n$, i.e., $\lim\limits_{i} L(f_i) = \int g\,dL$.

The assertion (b) has a similar proof.

In the sequel, for any $v \in V$, any $\mu \in [CV_0(X)]^*$, we shall use the notations: $B_v^0 = \left\{\mu \in [CV_0(X)]^*; \ |\mu|\left(\frac{1}{v}\right) \leq 1\right\}$, and $\sigma(\mu)$ for the support of μ.

Also, for any linear subspace $\mathcal{W} \subset CV_0(X)$ we denote by $\mathcal{W}^0$ its polar set, i.e.,

$$\mathcal{W}^0 = \left\{\mu \in [CV_0(X)]^*; \ \mu(w) = 0, \ \forall w \in \mathcal{W}\right\},$$

and for any convex set $S \subset [CV_0(X)]^*$ we denote by $Ext(S)$ the set of all extreme points of S.   □

Theorem 2.3.16 *If* $\mathcal{W} \subset CV_0(X)$ *is a linear subspace, then the closure of* $\mathcal{W}$ *in* $(CV_0(X), \omega_V)$ *is given by*

$$\overline{\mathcal{W}} = \left\{f \in CV_0(X); \ f\,|\sigma(\mu) \in \overline{\mathcal{W}\,|\sigma(\mu)}, \ \forall \mu \in Ext\left(B_v^0 \cap \mathcal{W}^0\right), \ \forall v \in V\right\}.$$

Proof We show only that for any function $g \in CV_0(X) \backslash \overline{\mathcal{W}}$ there exists $v \in V$ and $\mu \in Ext\left(B_v^{\,0} \cap \mathcal{W}^0\right)$ such that $g \,|\sigma(\mu) \notin \overline{\mathcal{W}} \,|\sigma(\mu)$. Indeed, using Hahn–Banach separation theorem, there exists $\lambda \in [CV_0(X)]^*$ such that $\lambda \in \mathcal{W}^0$ and $\lambda(g) \neq 0$. Let $v \in V$ be such that $|\lambda(f)| \leq p_v(f)$, $\forall f \in CV_0(X)$, i.e., $|\lambda| \left(\frac{1}{v}\right) \leq 1$. Hence $\lambda \in B_v^{\,0} \cap \mathcal{W}^0$. Since $B_v^{\,0} \cap \mathcal{W}^0$ is a compact convex subset of $[CV_0(X)]^*$ with respect to the weak topology and $\lambda(g) \neq 0$, it follows from Krein–Milman theorem that there exists $\mu \in Ext\left(B_v^{\,0} \cap W^0\right)$ such that $\mu(g) \neq 0$. Since $\mu \in \mathcal{W}^0$ we deduce, using Lemma 2.3.15, that $\int \varphi d\mu = 0$ for any $\varphi \in \overline{\mathcal{W}} \,|\sigma(\mu)$. Hence $g \,\big|\sigma(\mu) \notin \overline{\mathcal{W}} \,|\sigma(\mu)$. □

Corollary 2.3.17 *Let $\mathcal{W} \subset CV_0(X)$ be a linear subspace, and let $(P_i)_{i \in I}$ be a partition of X such that for any $v \in V$ and any $\mu \in Ext\left(B_v^{\,0} \cap W^0\right)$ there exists P_{i_μ} such that $\sigma(\mu) \subset P_{i_\mu}$. Then, we have*

$$\overline{\mathcal{W}} = \left\{f \in CV_0(X); \ f \,|P_i \in \overline{\mathcal{W} \,|P_i}\,, \ \forall i \in I\right\}.$$

The following result is a generalization of de Brange's lemma.

Lemma 2.3.18 *Let $\mathcal{W} \subset CV_0(X)$ be a linear subspace, $\mu \in Ext\left(B_v^{\,0} \cap \mathcal{W}^0\right)$ for some $v \in V$, and let f be a real valued continuous and bounded function on $\sigma(\mu)$ such that $\mu(f \cdot w) = 0$, $\forall w \in \mathcal{W}$. Then, f is constant on $\sigma(\mu)$.*

Proof Let $n, m \in \mathbb{N}^*$ be sufficiently large such that $\frac{1}{m} \cdot f < 1$; $0 < \frac{1}{n} \cdot \left(1 - \frac{1}{m} \cdot f\right)$ on $\sigma(\mu)$. Obviously, the function $g = \frac{1}{n} \cdot \left(1 - \frac{1}{m} \cdot f\right)$ has the same properties like f but $0 < g < 1$ on $\sigma(\mu)$.

We also denote by g the positive Borel extension of g on X such that $g = 0$ on $X \backslash \sigma(\mu)$. We consider the Radon measures μ_1, μ_2 on X given by

$$\mu_1 = \frac{g \cdot \mu}{|\mu| \left(\frac{g}{v}\right)}, \ \mu_2 = \frac{(1-g) \cdot \mu}{|\mu| \left(\frac{1-g}{v}\right)}.$$

Using Lemma 2.2.4 and Theorem 2.2.5, we have for any Radon measure λ on X:

$$\|\lambda\|_v = \sup\left\{|\lambda(g)|\,; \ g \in \mathcal{K}(X), \ p_v(g) \leq 1\right\} = |\lambda| \left(\frac{1}{v}\right), \text{ and}$$

$$B_v^{\,0} = \left\{\lambda \in \mathcal{M}(X); \ |\lambda| \left(\frac{1}{v}\right) \leq 1\right\}.$$

Particularly, we have $|\mu| \left(\frac{1}{v}\right) = \|\mu\|_v = 1$ since $\mu \in Ext\left(B_v^{\,0} \cap \mathcal{W}^0\right)$.

Furthermore, we have

$$\|\mu_1\|_v = |\mu_1|\left(\frac{1}{v}\right) = \frac{1}{|\mu|\left(\frac{g}{v}\right)} \cdot |g| \cdot |\mu|\left(\frac{1}{v}\right) = \frac{|\mu|\left(\frac{g}{v}\right)}{|\mu|\left(\frac{g}{v}\right)} = 1,$$

and similarly $\|\mu_2\|_v = 1$. On the other hand, we have $\mu_1(w) = \mu_2(w) = 0$ for all $w \in W$ and therefore $\mu_1, \mu_2 \in B_v^0 \cap W^0$. If we denote by $\alpha = |\mu|\left(\frac{g}{v}\right)$ and $\beta = |\mu|\left(\frac{1-g}{v}\right)$, we have $\alpha + \beta = 1$ and $\alpha \cdot \mu_1 + \beta \cdot \mu_2 = g \cdot \mu + (1 - g) \cdot \mu = \mu$. Since μ is an extreme point of $B_v^0 \cap W^0$, it follows that $\mu_1 = \mu_2 = \mu$. Hence $\frac{g}{\alpha} \cdot \mu = \mu$ and so $g = \alpha$ on $\sigma(\mu)$. Therefore, the function $f = m \cdot (1 - n \cdot g)$ is constant on $\sigma(\mu)$, and so the proof is finished. □

Definition 2.3.19 Let $\mathcal{A} \subset C(X, \mathbb{C})$ be an algebra with unit. A subset $S \subset X$ is called antisymmetric with respect to $\mathcal{A}$ if any $a \in \mathcal{A}$ which is real on S is constant on S.

We denote by $\mathcal{S}$ the family of all subsets of X which are antisymmetric with respect to $\mathcal{A}$. Obviously, $\mathcal{S} \neq \phi$ because for any $x \in X$, the set $\{x\} \in \mathcal{S}$.

Remark 2.3.20 The family $\mathcal{S}$ has the following properties:

(i) If $S_i \in \mathcal{S}$, $i = 1, 2$, and $S_1 \cap S_2 \neq \phi$, then $S_1 \cup S_2 \in \mathcal{S}$.
(ii) The closure $\overline{S}$ of any $S \in \mathcal{S}$ belongs to $\mathcal{S}$.
(iii) Any element $x \in X$ belongs to a maximal (with respect to the inclusion order relation) element of $\mathcal{S}$ denoted by S_x.
(iv) For any $x, y \in X$ we have either one or the other of the relations:

$$S_x = S_y, \quad S_x \cap S_y = \phi.$$

$$(v) X = \bigcup_{x \in X} S_x.$$

Theorem 2.3.21 *Let $\mathcal{A}$ be a subalgebra of $C(X, \mathbb{C})$ such that any element $a \in \mathcal{A}$ is a bounded function on the set $[v > 0]$ for each $v \in V$. If $W \subset CV_0(X)$ is a linear subspace with the property $\mathcal{A} \cdot W \subset W$, then the closure of W in $(CV_0(X), \omega_V)$ is given by*

$$\overline{W} = \left\{ f \in CV_0(X); \ f \mid S_x \in \overline{W \mid S_x}, \ \forall x \in X \right\}.$$

Proof First, we show that for any $v \in V$ and any extreme element $\mu \in Ext\left(B_v^0 \cap W^0\right)$ the set $\sigma(\mu)$ is antisymmetric with respect to $\mathcal{A}$. Indeed, since $1 = \|\mu\|_v = |\mu|\left(\frac{1}{v}\right)$, we deduce that $|\mu|([v = 0]) = 0$. Since any element $a \in \mathcal{A}$ is bounded on the set $[v > 0]$, then a is bounded on the closure $\overline{[v > 0]}$ of this set. Since $v = 0$ on $X\backslash[v > 0]$, we get

$$|\mu|\left(X\backslash\overline{[v > 0]}\right) = 0, \ \sigma(\mu) \subset \overline{[v > 0]},$$

and therefore any function $a \in \mathcal{A}$ is bounded on $\sigma(\mu)$. We have $\mu \in \mathcal{W}^0$ and $\mu(a \cdot w) = 0$ for any $a \in \mathcal{A}$ and any $w \in \mathcal{W}$. Using Lemma 2.3.18, we deduce that any element $a \in \mathcal{A}$ which is real on $\sigma(\mu)$ is constant on $\sigma(\mu)$. Therefore $\sigma(\mu) \in \mathcal{S}$, and so there exists $x_\mu \in X$ such that $\sigma(\mu) \subset S_{x_\mu}$.   □

Remark 2.3.22 If $\mathcal{A} \subset C(X, \mathbb{C})$ is a self-adjoint algebra, then any antisymmetric subset with respect to $\mathcal{A}$ is a set of constancy for $\mathcal{A}$. Particularly, for any $x \in X$ we have

$$S_x = [x]_{\mathcal{A}} = \{y \in X; \ a(y) = a(x), \ \forall a \in \mathcal{A}\}.$$

Indeed, any element $a \in \mathcal{A}$ is of the form $a = a' + i \cdot a''$ where a', a'' are real functions on X. Since $a' = \frac{a+\bar{a}}{2} \in \mathcal{A}$, $a'' = \frac{a-\bar{a}}{2 \cdot i} \in \mathcal{A}$, we deduce that a' and a'' are constant on any antisymmetric set with respect to $\mathcal{A}$, and therefore a is constant on any such set. From the previous remark, Theorem 2.3.21 becames a generalization of Theorem 2.3.9.

Definition 2.3.23 Let $\mathcal{M} \subset C(X, \mathbb{R})$ and $\mathcal{W} \subset CV_0(X)$ be two nonempty subsets. A subset S of X will be called antialgebraic with respect to the pair $(\mathcal{M}, \mathcal{W})$ if any element $m \in \mathcal{M}$ such that

$$m \cdot w \, |S \in \mathcal{W}\, |S, \ \forall w \in \mathcal{W}$$

is constant on S.

If we denote by $\mathcal{T}$ the family of all antialgebraic subsets of X with respect to the pair$(\mathcal{M}, \mathcal{W})$, then for any $x \in X$ the singleton $\{x\}$ belongs to $\mathcal{T}$.

The set $\mathcal{T}$ endowed with the inclusion order relation has similar properties as family $\mathcal{S}$ in the Remark 2.3.20. For any $x \in X$, we denote by T_x the maximal $(\mathcal{M}, \mathcal{W})$ −antialgebraic subset containing x. We have $X = \bigcup_{x \in X} T_x$, and $\{T_x\}_{x \in X}$ is a partition of X.

Remark 2.3.24 If we have two pairs$(\mathcal{M}_1, \mathcal{W}_1)$, $(\mathcal{M}_2, \mathcal{W}_2)$ as above, and we denote by $\mathcal{T}_i$ the family of all antialgebraic subsets of X with respect to $(\mathcal{M}_i, \mathcal{W}_i)$, $i = 1, 2$, then we have

$$\mathcal{M}_1 \subset \mathcal{M}_2 \ \Rightarrow \ \mathcal{T}_2 \subset \mathcal{T}_1 \ \Rightarrow \ T_{2x} \subset T_{1x}, \ \forall x \in X,$$

where, for any $x \in X$, T_{ix} denotes the maximal element from $\mathcal{T}_i$ containing the point x.

Theorem 2.3.25 *Let $\mathcal{M}$ be a V-bounded (i.e., any function $m \in \mathcal{M}$ is bounded on the support of any weight $v \in V$) nonempty subset of $C(X, \mathbb{R})$, and let $\mathcal{W}$ be a linear subspace of $CV_0(X)$. If we denote by T_x the maximal antialgebraic set with respect to pair $(\mathcal{M}, \mathcal{W})$ such that $x \in T_x$, then we have*

$$\overline{\mathcal{W}} = \left\{ f \in CV_0(X); \ f \left| T_x \in \overline{\mathcal{W} | T_x}, \ \forall x \in X \right. \right\}.$$

Proof Applying Corollary 2.3.17 will be sufficient to show that for any $v \in V$ and any $\mu \in Ext\left(B_v{}^0 \cap \mathcal{W}^0\right)$ the set $\sigma(\mu)$ is included in T_{x_0} for some $x_0 \in X$, or equivalently to show that $\sigma(\mu)$ is an antialgebraic set with respect to the pair $(\mathcal{M}, \mathcal{W})$. Let $m \in \mathcal{M}$ be such that $m \cdot w \, |\sigma(\mu) \in \mathcal{W} \, |\sigma(\mu)$ for all $w \in \mathcal{W}$. Using Lemma 2.3.18, we deduce that m is constant on $\sigma(\mu)$. Hence $\sigma(\mu)$ is an antialgebraic set with respect to the pair $(\mathcal{M}, \mathcal{W})$ and the proof is finished. $\qquad\square$

Theorem 2.3.26 *Let $\mathcal{W}$ be a linear subspace of $CV_0(X)$ and let $\mathcal{M}$ be a subset of continuous real bounded functions on X. If for any $x \in X$ we denote by T_x the maximal antialgebraic subset with respect to pair $(\mathcal{M}, \mathcal{W})$ containing x, then we have*

$$\overline{\mathcal{W}} = \left\{ f \in CV_0(X); \ f \left| T_x \in \overline{\mathcal{W} | T_x}, \ \forall x \in X \right. \right\}.$$

Proof The assertion follows from Theorem 2.3.25 since any function of $\mathcal{M}$ is V-bounded. $\qquad\square$

Corollary 2.3.27 *Let $\mathcal{M}, \ \mathcal{W}$ be as in Theorem 2.3.26. If we suppose in addition that $\mathcal{M} \cdot \mathcal{W} \subset \mathcal{W}$, then we have*

$$\overline{\mathcal{W}} = \left\{ f \in CV_0(X); \ f \left| [x]_{\mathcal{M}} \in \overline{\mathcal{W} | [x]_{\mathcal{M}}}, \ \forall x \in X \right. \right\},$$

where, for any $x \in X$, we have denoted

$$[x]_{\mathcal{M}} = \{ y \in X; \ m(y) = m(x), \ \forall m \in \mathcal{M} \}.$$

Proof From the hypothesis $\mathcal{M} \cdot \mathcal{W} \subset \mathcal{W}$, we deduce that $[x]_{\mathcal{M}} = T_x$ where T_x is the maximal $(\mathcal{M}, \mathcal{W})$-antialgebraic set containing x. The proof is finished applying Theorem 2.3.26. $\qquad\square$

2.4   Stone–Weierstrass Theorem for Convex Cones in a Weighted Space

In this section, we consider a convex cone $C \subset CV_0(X, \mathbb{R})$ and we denote by C^0 its polar set, i.e.,

$$C^0 = \left\{ \mu \in [CV_0(X, \mathbb{R})]^*; \ \mu(f) \leq 0, \ \forall f \in C \right\}.$$

For any weight $v \in V$, we denote by $Ext\left\{ B_v{}^0 \cap C^0 \right\}$ the set of all extreme points of the compact convex set $B_v{}^0 \cap C^0$ of the dual $[CV_0(X, \mathbb{R})]^*$ of the locally convex space $(CV_0(X, \mathbb{R}), \omega_V)$. We remember that the closure of any convex cone in an arbitrary locally convex space coincides with its bipolar with respect to the natural duality. Hence we have

$$\overline{C} = \left\{ f \in CV_0(X); \ \mu(f) \leq 0, \ \forall \mu \in C^0 \right\}.$$

The following result is a generalization of de Brange's Lemma for a convex cone.

Lemma 2.4.1 *Let $C \subset CV_0(X, \mathbb{R})$ be a convex cone, $v \in V$ and $\mu \in Ext\left\{ B_v{}^0 \cap C^0 \right\}$. If $\sigma(\mu)$ denotes the support of the Radon measure μ, then any function $\varphi \in C(X, \mathbb{R})$ such that*

$$(i) \ \ 0 \leq \varphi(x) \leq 1, \ \forall x \in \sigma(\mu),$$

$$(ii) \ \ \varphi \cdot f \, |\sigma(\mu), \ (1 - \varphi) \cdot f \, |\sigma(\mu) \in \overline{C} \, |\sigma(\mu), \ \forall f \in C$$

is a constant function on $\sigma(\mu)$.

Proof Since $\mu \in Ext\left\{ B_v{}^0 \cap C^0 \right\}$, we deduce that $\|\mu\|_v = |\mu|\left(\frac{1}{v}\right) = 1$. If $|\mu|(\varphi) = 0$ or $|\mu|(1 - \varphi) = 0$ we have $\varphi = 0$ or $\varphi = 1$ on $\sigma(\mu)$.

We suppose now that $|\mu|(\varphi) \neq 0$ and $|\mu|(1 - \varphi) \neq 0$ and we consider the measures μ_1, μ_2 given by

$$\mu_1 = \frac{\varphi \cdot \mu}{|\mu|\left(\frac{\varphi}{v}\right)}, \quad \mu_2 = \frac{(1 - \varphi) \cdot \mu}{|\mu|\left(\frac{1 - \varphi}{v}\right)}.$$

Furthermore, we have

$$\|\mu_1\|_v = |\mu_1|\left(\frac{1}{v}\right) = \frac{|\mu|\left(\varphi \cdot \frac{1}{v}\right)}{|\mu|\left(\frac{\varphi}{v}\right)} = 1, \quad \|\mu_2\|_v = |\mu_2|\left(\frac{1}{v}\right) = \frac{|\mu|\left((1 - \varphi) \cdot \frac{1}{v}\right)}{|\mu|\left(\frac{1 - \varphi}{v}\right)} = 1.$$

Since $\varphi \cdot h \,|\sigma(\mu) \in \overline{C} \,|\sigma(\mu)$ and $(1 - \varphi) \cdot h \,|\sigma(\mu) \in \overline{C} \,|\sigma(\mu)$, $\forall h \in C$ and using Lemma 2.3.15, having in mind that $\mu \in C^0$, we deduce

$$\mu_1(h) = \frac{\mu(\varphi \cdot h)}{|\mu|\left(\frac{\varphi}{v}\right)} \leq 0, \quad \mu_2(h) = \frac{\mu[(1 - \varphi) \cdot h]}{|\mu|\left(\frac{1-\varphi}{v}\right)} \leq 0.$$

Hence $\mu_1, \mu_2 \in Ext\left\{B_v{}^0 \cap C^0\right\}$. On the other hand, since

$$|\mu|\left(\frac{\varphi}{v}\right) \cdot \mu_1 + |\mu|\left(\frac{1-\varphi}{v}\right) \cdot \mu_2$$

$$= \mu \; and \; |\mu|\left(\frac{\varphi}{v}\right) + |\mu|\left(\frac{1-\varphi}{v}\right) = 1,$$

we get $\mu_1 = \mu_2 = \mu$, i.e.,

$$\frac{\varphi}{|\mu|\left(\frac{\varphi}{v}\right)} = 1 \; \text{ on } \; \sigma(\mu). \qquad \square$$

Theorem 2.4.2 *A function $f \in CV_0(X)$ belongs to the closure $\overline{C}$ of the convex cone C in the locally convex space $(CV_0(X, \mathbb{R}, \; \omega_V)$ if and only if for any weight $v \in V$ and any $\mu \in Ext\left\{B_v{}^0 \cap C^0\right\}$ we have*

$$\mu(f) \leq 0.$$

Proof We show only that if $f \in CV_0(X, \mathbb{R}) \backslash \overline{C}$ there exist $v \in V$ and $\mu \in Ext\left\{B_v{}^0 \cap C^0\right\}$ such that $\mu(f) > 0$.

Indeed, if $f \notin \overline{C} = C^0$ there exists $\lambda \in [CV_0(X, \mathbb{R})]^*$, $\lambda \in C^0$ such that $\lambda(f) > 0$. Let us consider $v \in V$ such that $|\lambda|\left(\frac{1}{v}\right) = 1 = \|\lambda\|_v$, $\lambda \in B_v{}^0$. Since the following map:

$$\theta : B_v{}^0 \cap C^0 \to \mathbb{R}, \; \theta(f) = \int f d\lambda,$$

is a continuous affine function, if we endow the set $B_v{}^0 \cap C^0$ with a trace of the weak topology on $[CV_0(X, \mathbb{R})]^*$ and the maximum of this map is realized on a point $\mu \in Ext\left\{B_v{}^0 \cap C^0\right\}$, we deduce that $\mu(f) \geq \lambda(f) > 0$.

The following statement of a convex cone is a procedure to describe the closure of a convex cone in some circumstances. $\qquad \square$

Corollary 2.4.3 *Let $C \subset CV_0(X, \mathbb{R})$ be a convex cone and let $(P_\alpha)_{\alpha \in I}$ be a partition of X such that for any $v \in V$ and any $\mu \in Ext \left\{ B_v^0 \cap C^0 \right\}$ there exists $\alpha \in I$ such that the support of μ, $\sigma(\mu) \subset P_\alpha$.*

Then, we have

$$\overline{C} = \left\{ f \in CV_0(X, \mathbb{R}); \ f \big| P_\alpha \in \overline{C \, | P_\alpha}, \ \forall \alpha \in I \right\}.$$

Now, we state such kind of circumstances.

Let $\mathcal{M}$ be a nonempty subset of $C(X, [0, 1])$ such that, if $\varphi \in \mathcal{M}$, then its "complement" $1 - \varphi \in \mathcal{M}$.

Definition 2.4.4 A subset $S \subset X$ is called antialgebraic with respect to the pair $(\mathcal{M}, C)$ if any function $\varphi \in \mathcal{M}$ with the properties:

$$\varphi \cdot f \big| S \in \overline{C \, | S} \ , \ (1-\varphi) \cdot f \big| S \in \overline{C \, | S} \ , \ \forall f \in C,$$

is a constant function on S.

Furthermore, we denote by $\mathcal{B}$ the family of all subsets of X antialgebraic with respect to the pair $(\mathcal{M}, C)$.

The following assertions are almost obvious.

  (i) $\{x\} \in \mathcal{B}, \ \forall x \in X$
 (ii) $B_1, B_2 \in \mathcal{B}, \ B_1 \cap B_2 \neq \phi \ \Rightarrow B_1 \cup B_2 \in \mathcal{B}$
(iii) $B \in \mathcal{B} \Rightarrow \overline{B} \in \mathcal{B}$
 (iv) For any upper directed family $(B_\alpha)_{\alpha \in I}$ from $\mathcal{B}$ we have $\bigcup_{\alpha \in I} B_\alpha \in \mathcal{B}$.
  (v) For any $x \in X$ we denote $B_x = \cup \{B; \ B \in \mathcal{B}, \ x \in B\}$. We have

$$B_x = \overline{B_x} \in \mathcal{B}, \ B_x \cap B_y = \phi \ \text{ if } \ B_x \neq B_y.$$

The family $(B_x)_{x \in X}$ is a partition of X, and for any $B \in \mathcal{B}$ there exists $x \in X$ such that $B \subset B_x$.

Remark 2.4.5 With the above notations, we have

For any $v \in V$ and any $\mu \in Ext \left\{ B_v^0 \cap C^0 \right\}$ the set $\sigma(\mu)$ belongs to $\mathcal{B}$.

Indeed, if $\varphi \in \mathcal{M}$ is that $\varphi \cdot f \big| \sigma(\mu) \in \overline{C \, | \sigma(\mu)} \ , \ (1-\varphi) \cdot f \big| \sigma(\mu) \in \overline{C \, | \sigma(\mu)} \ , \ \forall f \in C,$ we deduce, from Lemma 2.4.1, that φ is a constant function on $\sigma(\mu)$.

Theorem 2.4.6 *Let X be a locally compact Hausdorff space, V be a Nachbin family of weights on X, $C \subset CV_0(X, \mathrm{R})$ be a convex cone and $\mathcal{M} \subset C(X, [0, 1])$ be a nonempty subset with complement. Then*

$$\overline{C} = \left\{ f \in CV_0(X, \mathrm{R}); \ f \,\middle|\, B_x \in \overline{C \,|\, B_x}, \ \forall x \in X \right\},$$

where $(B_x)_{x \in X}$ is the family of all maximal subsets of X antialgebraic with respect to the pair $(\mathcal{M}, C)$.

Proof The assertion follows from the above remark and from the Corollary 2.4.3.   □

Corollary 2.4.7 *Let $X, V, C, \mathcal{M}$ be as in Theorem 2.4.6 such that $\mathcal{M} \cdot C \subset C$. Then, we have*

$$\overline{C} = \left\{ f \in CV_0(X, \mathbb{R}); \ f \,\middle|\, [x]_\mathcal{M} \in \overline{C \,|\, [x]_\mathcal{M}}, \ \forall x \in X \right\},$$

where, for any $x \in X$, $[x]_\mathcal{M} = \{y \in X; \ m(y) = m(x), \ \forall m \in \mathcal{M}\}$.

Proof Since $\mathcal{M}$ is a set with complement and $\mathcal{M} \cdot C \subset C$, we deduce that all functions from $\mathcal{M}$ are constant on any antialgebraic set with respect to the pair $(\mathcal{M}, C)$, and for any $x \in X$ the set $[x]_\mathcal{M}$ is an antialgebraic set with respect to the pair $(\mathcal{M}, C)$. Hence the set $[x]_\mathcal{M} = B_x$, $\forall x \in X$. The assertion follows now from Theorem 2.4.6.   □

Corollary 2.4.8 *Let $X, V, C, \mathcal{M}$ be as in Corollary 2.4.7. Moreover, we suppose that $\mathcal{M}$ separates the points of X, i.e., for any $x, y \in X$, $x \neq y$ there exists $m \in \mathcal{M}$ such that $m(x) \neq m(y)$. Let us denote:*

$$X_- = \{x \in X; \ f(x) \leq 0, \ \forall f \in C\}$$

$$X_+ = \{x \in X; \ f(x) \geq 0, \ \forall f \in C\}$$

Then we have:

$$\overline{C} = \{f \in CV_0(X, \mathbb{R}); \ f \geq 0 \text{ on } X_+, \ f \leq 0 \text{ on } X_-\}$$

Example 2.4.9 Let $X = \mathbb{R}$ and let $\varphi : \mathbb{R} \rightarrow (0, 1)$ be the strictly continuous homeomorphism given by $\varphi(x) = \frac{2}{\pi} \cdot arctg(e^x)$. On the space $C_b(\mathbb{R})$ of all real bounded and continuous functions on $\mathbb{R}$ we consider the strict topology β given by the family of seminorms:

$$f \rightarrow p_g(f) = \sup_{x \in \mathrm{R}} g(x) \cdot |f(x)|, \ \forall f \in C_b(\mathbb{R}),$$

where g runs the set $C_0^+(\mathbb{R})$ of all positive, continuous functions on $\mathbb{R}$ vanishing at infinity. In fact, $\beta = \omega_V$ where the Nachbin family of weights V is just $C_0^+(\mathbb{R})$. We know that

$$CV_0(\mathbb{R}) = C_b(\mathbb{R}).$$

Let us consider the convex cone $\mathcal{C}$ in $C_b^+(\mathbb{R})$ given by

$$\mathcal{C} = \left\{ P(\varphi, 1 - \varphi);\ P(x, y) = \sum_{i,j=1}^{n} a_{ij} \cdot x^i \cdot y^j,\ a_{ij} \geq 0,\ n \in \mathbb{N} \right\}.$$

Obviously, $\mathcal{C}$ separates the points of X since $\varphi \in \mathcal{C}$. Using the notations from Corollary 2.4.8, we have $X_0 = \phi$, $X_+ = \mathbb{R}$, and therefore $\overline{\mathcal{C}} = C_b^+(\mathbb{R})$, i.e., for any $f \in C_b^+(\mathbb{R})$ and any $g \in C_0^+(\mathbb{R})$ there exists a sequence $(P_k)_k$ of polynomials $P_k(x, y) = \sum_{i,j=1}^{n_k} a_{ij}{}^k \cdot x^i \cdot y^j$, $a_{ij}{}^k \geq 0$ such that the sequence $(g \cdot P_k(\varphi, 1 - \varphi))_k$ converges uniformly to $g \cdot f$ on $\mathbb{R}$.

2.5 Approximation of Vector Valued Functions

Many of the results presented before may be extended to the sets of vector valued functions. We give, essentially, one result in this direction.

Let X be a locally compact Hausdorff space, let E be a locally convex space, and let $\mathscr{P}$ be the family of seminorms of E. We denote by $C(X, E)$ the set of all continuous functions $f : X \to E$ and we denote by $C_0(X, E)$ the subset of continuous functions vanishing at infinity and by $\mathcal{K}(X, E)$ the subset of continuous functions with compact support. For instance, the function $f \in C(X, E)$ belongs to $C_0(X, E)$ if for any $\varepsilon \in \mathbb{R}$, $\varepsilon > 0$, and any $p \in \mathscr{P}$ there exists a compact subset $K_{\varepsilon, p}$ of X such that

$$p[f(x)] < \varepsilon,\ \forall x \in X \backslash K_{\varepsilon, p}.$$

If V is a Nachbin family of weights on X, $V \subset C^+(X, \mathbb{R})$, we denote

$$CV_0(X, E) = \{ f \in C(X, E);\ v \cdot f \in C_0(X, E),\ \forall v \in V \}.$$

We endow this linear space with the weighted topology $\omega_{V, P}$ given by the family of seminorms $\| \| \|_{v, p}$ defined by

$$\| f \|_{v, p} = \sup \{ v(x) \cdot p[f(x)],\ \forall x \in X \},\ \forall f \in CV_0(X, E).$$

A base of neighbourhoods of the origin of $CV_0(X, E)$ is the family $(B_{v,p})_{v \in V, p \in P}$ given by

$$B_{v,p} = \{f \in CV_0(X, E);\ \|f\|_{v,p} \le 1\}.$$

As in the scalar case, one can see that $\mathcal{K}(X, E)$ is a dense subset of $CV_0(X, E)$ with respect to the weighted topology $\omega_{V,P}$.

For any $p \in \mathscr{P}$ and any $f \in \mathcal{K}(X, E)$, we denote

$$\|f\|_p = \sup_{x \in X} p[f(x)].$$

Obviously, $\|f\|_p < \infty$ since $p : E \to \mathbb{R}_+$ is a continuous function on the locally compact space E, and $f(X) = f(K_f) \cup \{0\}$ is a compact subset of E, where K_f denoted the support of f. If we endow $\mathcal{K}(X, E)$ with the family of seminorms $(\|\|_p)_{p \in P}$, then $\mathcal{K}(X, E)$ becomes a locally convex space, and we shall denote by τ the topology given by these seminorms $(\|\|_p)_{p \in P}$.

For any compact subset $K \subset X$, we denote by $\mathcal{K}(K, E)$ the set of all functions $f :$ $X \to E$ which vanishes outside K.

Definition 2.5.1 A linear map $U : \mathcal{K}(X, E) \to \mathbb{R}$ is called a Radon measure if for any compact subset $K \subset X$ the restriction of U to $\mathcal{K}(K, E)$ is continuous if we endow $\mathcal{K}(K, E)$ with the trace of τ on $\mathcal{K}(K, E)$, i.e., there exists $p \in P$ and $\alpha \in \mathbb{R}_+$ such that $|U(f)| \le \alpha \cdot \|f\|_p, \forall f \in \mathcal{K}(K, E)$.

Generally, the number α and the seminorm p change if we change the compact K of X. If α and p are independent from K, we say that U is a p-bounded or p-dominated Radon measure, i.e.,

$$|U(f)| \le \alpha \cdot \|f\|_p, \ \forall f \in \mathcal{K}(X, E).$$

The smallest $\alpha \in \mathbb{R}_+$ in the previous inequality will be denoted by $\|U\|_p$. In fact,

$$\|U\|_p = \sup \{|U(f)|;\ f \in \mathcal{K}(X, E),\ p(f) \le 1\}.$$

We shall denote by $M_b(X, E)$ the set of all Radon measures U on $\mathcal{K}(X, E)$ for which there exists $p \in \mathscr{P}$ (p depending on U) such that U is p-dominated. For any $p \in \mathscr{P}$ we denote by $B_{1,p}^0$ or simply B_p^0 the set $\{U \in M_b(X, E);\ \|U\|_p \le 1\}$.

Obviously, $M_b(X, E)$ is the dual of the locally convex space $(\mathcal{K}(X, E),\ \tau)$. From Alaoglu's theorem, the set B_p^0 is a compact convex subset of $M_b(X, E)$ with respect to the weak topology $\sigma (M_b(X, E),\ \mathcal{K}(X, E))$.

Remark 2.5.2 If the Radon measure $U : \mathcal{K}(X, E) \to \mathbb{R}$ is dominated by p, then there exists a smallest positive Radon measure μ on X such that

$$|U(f)| \leq \int p[f(x)]d\mu(x), \; \forall f \in \mathcal{K}(X, E).$$

This measure will be denoted by $|U|$. We have $\|U\|_p = |U|(1)$.

In fact, for any continuous, positive function φ on X vanishing outside a compact subset of X, we have

$$|U|(\varphi) = \int \varphi d|U| = \sup\{U(f); \; f \in \mathcal{K}(X, E), \; p \circ f \leq \varphi\}.$$

We know that the subset $\mathcal{K}(X, \mathbb{R})$ is dense in the linear space $\mathcal{L}^1(|U|)$ of all numerical functions on X which are integrable with respect to the positive measure $|U|$ on X, endowed with seminorm $\|f\|_1 = \int |f| d|U|$, i.e., for any $f \in \mathcal{L}^1(|U|)$ there exists a sequence $(\varphi_n)_n$ in $\mathcal{K}(X, \mathbb{R})$ which converges $|U|$-a.e. on X to f and such that

$$\lim_{n,m \to \infty} \int |\varphi_n - \varphi_m| d|U| = 0.$$

Let us denote by $\mathcal{L}^1(U, E)$ the set of all functions $f : X \to E$ such that there exists a sequence $(f_n)_n$ in $\mathcal{K}(X, E)$ with the properties:

$$\lim_{n \to \infty} p[f_n(x) - f(x)] = 0, \; \forall x \in X \backslash A, \; |U|(A) = 0, \; \text{and}$$

$$\lim_{n,m \to \infty} |U|[p(f_n - f_m)] = \lim_{n,m \to \infty} \int p(f_n - f_m)(x)d|U|(x) = 0.$$

Since we have $|U(f_n - f_m)| \leq |U|[p(f_n - f_m)]$, we deduce that the sequence $(U(f_n))_n$ is convergent. We denote by $U(f)$ or $\int f dU$ the limit of this sequence.

One can show that this limit doesn't depend on the sequence $(f_n)_n$ having the above properties, and we have

$$|U(f_n) - U(f)| = |U(f_n - f)| \leq \int p(f_n - f)d|U|, \; \lim_{n \to \infty} \int p(f_n - f)d|U| = 0.$$

The basic argument here is the following assertion:

If $(f_n)_n$ is a sequence in $\mathcal{K}(X, E)$ such that $\lim_{n,m \to \infty} |U|[p(f_n - f_m)] = 0$ and $\lim_{n \to \infty} p[f_n(x)] = 0$, $|U|$-a.e. on X, then $\lim_{n \to \infty} |U|[p(f_n)] = 0$ and $\lim_{n \to \infty} U(f_n) = 0$

Theorem 2.5.3 *Let U be a p-bounded Radon measure on $\mathcal{K}(X, E)$, and let f, (resp. g) be two elements from $\mathcal{L}^1(U, E)$ (resp. $\mathcal{L}^1(|U|)$) such that one of the positive functions $p \circ f$ or g is bounded on X, $|U|$-a.e. Then the function $x \to f(x) \cdot g(x)$ belongs to $\mathcal{L}^1(U, E)$. If $g \in \mathcal{L}^1(|U|)$ or g is bounded $|U|$-a.e. on X, then the map on $\mathcal{K}(X, E)$:*

$$\varphi \to \int \varphi \cdot g dU$$

is a p-dominated Radon measure on $\mathcal{K}(X, E)$ denoted by $g \cdot U$, and we have

$$\|g \cdot U\|_p = \int |g| \, d \, |U|.$$

Proof Let $(f_n)_n$ (resp. $(g_n)_n$) be sequences in $\mathcal{K}(X, E)$, (resp. $\mathcal{K}(X, \mathbb{R})$) such that

$$\lim_{n,m \to \infty} \int p(f_n - f_m) d \, |U| = \lim_{n,m \to \infty} \int p(g_n - g_m) d \, |U| = 0$$

and

$$\lim_{n \to \infty} p(f_n - f)(x) = \lim_{n \to \infty} p(g_n - g)(x) = 0, \quad |U|\text{-a.e.}$$

We consider only the case where $g \in \mathcal{L}^1(|U|)$ and $p \circ f$ is bounded on X. We may suppose that $(p \circ f)(x) \le 1$ and $(p \circ f_n)(x) \le 2$ on X for any $n \in \mathbb{N}$. Furthermore, we have

$$\lim_{n \to \infty} p(f_n \cdot g_n - f \cdot g) \le \lim_{n \to \infty} p[f_n \cdot (g_n - g)] + \lim_{n \to \infty} p[(f_n - f) \cdot g]$$

$$= \lim_{n \to \infty} |g_n - g| \cdot p(f_n) + |g| \cdot p(f_n - f) = 0, \quad |U|\text{-a.e, and}$$

$$\int p(f_n \cdot g_n - f_m \cdot g_m) d \, |U| \le \left| \int |g_n - g_m| \cdot p(f_n) \right| d \, |U| + \int |g_m| \cdot p(f_n - f_m) d \, |U|.$$

Using Lebesgue domination theorem, we get

$$\lim_{n,m \to \infty} \int p(f_n \cdot g_n - f_m \cdot g_m) d \, |U| = 0,$$

and therefore $f \cdot g \in \mathcal{L}^1(U, E)$, $\lim_{n \to \infty} \int p(f \cdot g - f_n \cdot g_n) d \, |U| = 0$, and $U(f \cdot g) = \lim_{n \to \infty} U(f_n \cdot g_n)$.

Obviously, for any $f \in \mathcal{K}(X, E)$, we have

$$U(f \cdot g) = \lim_{n \to \infty} U(f \cdot g_n),$$

$$|U(f \cdot g)| = \lim_{n \to \infty} |U(f \cdot g_n)| \le \lim_{n \to \infty} |U| [p(f \cdot g_n)] \le \lim_{n \to \infty} \|f\|_p \cdot |U| (|g_n|) = \|f\|_p \cdot |U| (|g|).$$

Hence the map $g \cdot U : \mathcal{K}(X, E) \to \mathbb{R}$, given by

$$(g \cdot U)(f) = \int f \cdot g dU, \quad \forall f \in \mathcal{K}(X, E)$$

is a p-dominated Radon measure on $\mathcal{K}(X, E)$ and $|(g \cdot U)(f)| \le |U(|g|)| \cdot \|f\|_p$. In fact, we have

$$\|(g \cdot U)\|_p = |U|(|g|).$$

Furthermore, we give a characterization of the dual of locally convex space $(CV_0(X, E), \omega_{V,P})$. □

Theorem 2.5.4 *For any $p \in \mathscr{P}, v \in V$ and any linear map $L : CV_0(X, E) \to \mathbb{R}$, we have*

$$|L(f)| \le \alpha \cdot \|f\|_{v,p}, \quad \forall f \in CV_0(X, E),$$

if and only if there exists a Radon measure U on $\mathcal{K}(X, E)$, dominated by p, such that

$$L = v \cdot U \text{ or } L(\psi) = U(v \cdot \psi), \quad \forall \psi \in \mathcal{K}(X, E).$$

Proof We show only the if part of the assertion. Let $L \in B_{v,p}^0$ with $\|L\|_{v,p} = 1$, i.e., the smallest positive number α which satisfies the inequality:

$$|L(f)| \le \alpha \cdot \|f\|_{v,p}, \quad \forall f \in CV_0(X, E) \text{ or only } f \in \mathcal{K}(X, E),$$

is 1. We denote by $|L|$ the positive Radon measure on $\mathcal{K}(X, \mathbb{R})$ given by

$$|L|(\varphi) = \sup\{|L(\psi)|, \ \psi \in \mathcal{K}(X, E), \ p(\psi) \le \varphi\}, \quad \forall \varphi \in \mathcal{K}^+(X, E).$$

We have $|L(\psi)| \le |L|[p(\psi)]$ for all $\psi \in \mathcal{K}(X, E)$, and moreover $|L|\left(\frac{1}{v}\right) = 1 = \|L\|_{v,p}$.
　Indeed

$$|L|\left(\frac{1}{v}\right) = \sup\left\{|L|(\varphi); \ \varphi \in \mathcal{K}^+(X, \mathbb{R}), \ \varphi \le \frac{1}{v}\right\}$$

$$= \sup\left\{|L(\psi)|; \ \psi \in \mathcal{K}(X, E), \ p(\psi) \le \varphi \le \frac{1}{v}, \ \varphi \in \mathcal{K}^+(X, \mathbb{R})\right\}$$

$$= \sup\left\{|L(\psi)|; \ \psi \in \mathcal{K}(X, E), \ p(\psi) \le \frac{1}{v}\right\}$$

$$= \sup\{|L(\psi)|; \ \psi \in \mathcal{K}(X, E), \ p(\psi \cdot v) \le 1\} = \|L\|_{v,p} = 1.$$

Let us consider an increasing sequence $(\varphi_n)_n$ in $\mathcal{K}^+(X, \mathbb{R})$ such that the sequence $(|L|(\varphi_n))_n$ increases to $\left| L\left(\frac{1}{v}\right) \right| = 1$. Obviously, $\lim\limits_{n \to \infty} |L|(|\varphi_n - \varphi_m|) = 0$, and for any $\psi \in \mathcal{K}(X, E)$ we have

$$|L(\psi \cdot \varphi_n) - L(\psi \cdot \varphi_m)|$$
$$= |L[\psi \cdot (\varphi_n - \varphi_m)]| \leq\leq |L|(|\varphi_n - \varphi_m|$$
$$\cdot p(\psi) \leq |\psi|_p \cdot |L|(|\varphi_n - \varphi_m|).$$

Therefore, the sequence $(L(\psi \cdot \varphi_n))_n$ is convergent. If we denote by $\left(\frac{1}{v} \cdot L\right)(\psi)$ its limit, then the map $\psi \to \left(\frac{1}{v} \cdot L\right)(\psi)$ is a p-dominated Radon measure on $\mathcal{K}(X, E)$.

Indeed, for any $\psi \in \mathcal{K}(X, E)$ such that $\|\psi\|_p \leq 1$, or $p(\psi) \leq 1$ we have

$$|L(\psi \cdot \varphi_n)| \leq |L|[\varphi_n \cdot p(\psi)] \leq |L|(\varphi_n) \leq |L|\left(\frac{1}{v}\right),$$

i.e., the map $\frac{1}{v} \cdot L$ is p-dominated, and

$$\left\| \frac{1}{v} \cdot L \right\|_p = \left| \frac{1}{v} \cdot L \right|(1) = |L|\left(\frac{1}{v}\right) = 1.$$

If we denote $U = \frac{1}{v} \cdot L$, we have $L = v \cdot U$.   □

Remark 2.5.5 With the above notations for any $p \in \mathscr{P}$, $v \in V$, we have

$$B^0_{p,v} = \left\{ v \cdot U; \ U \in B^0_p \right\} = v \cdot B^0_p \text{ and}$$

$$\|v \cdot U\|_{p,v} = \|U\|_p.$$

If C is a convex cone in $CV_0(X, E)$, then the closure of C in the locally convex space $\left(CV_0(X, E), \omega_{V}.\right)$ coincides with the set C^0. A variant of de Brange's lemma may be stated in this context.

Theorem 2.5.6 *If $L \in B^0_{v,p} \cap C^0$, $L \neq 0$ is an extreme point of the compact convex set $B^0_{v,p} \cap C^0$ and $h : X \to [0, 1]$ is a continuous function such that $h \cdot c$, $(1 - h) \cdot c$ belongs to C for all $c \in C$, then h is constant on $\sigma(|L|)$ —the support of the positive measure $|L|$ defined as before:*

$$|L|(\varphi) = \sup\{|L(\psi)|, \ \psi \in \mathcal{K}(X, E), \ p(\psi) \leq \varphi\}, \ \forall \varphi \in \mathcal{K}^+(X, E).$$

Definition 2.5.7 If $\mathcal{M} \subset C(X, [0, 1])$ is such that for any $h \in \mathcal{M}$ the function $1 - h$ belongs to $\mathcal{M}$, then a subset $S \subset X$ is called antialgebraic with respect to the pair $(\mathcal{M}, \mathcal{C})$ if any element $h \in \mathcal{M}$ such that $h \cdot c\,|S \in \mathcal{C}\,|S$ and $(1 - h) \cdot c\,|S \in \mathcal{C}\,|S$ for any $c \in \mathcal{C}$ is a constant function on S.

Obviously, any singleton of X is an antialgebraic set with respect to any pair $(\mathcal{M}, \mathcal{C})$ and any singleton $\{x\}$ is included in a maximal $(\mathcal{M}, \mathcal{C})$-antialgebraic subset S_x and for any $x, y \in X, x \neq y$ we have either $S_x = S_y$, either $S_x \cap S_y = \phi$.

From Theorem 2.5.6, we deduce that the support $\sigma(L)$ of any extreme point L of $B^0_{v,p} \cap \mathcal{C}^0$ is antialgebraic with respect to any pair $(\mathcal{M}, \mathcal{C})$.

With similar arguments as in the scalar case, we have the following theorems.

Theorem 2.5.8 *If $\mathcal{C} \subset CV_0(X, E)$ is a convex cone, then the closure of $\mathcal{C}$ in $\left(CV_0(X, E), \omega_{V,P}\right)$ is given by*

$$\overline{\mathcal{C}} = \left\{ f \in CV_0(X, E); f|\sigma(L) \in \overline{\mathcal{C}|\sigma(L)}, \forall L \in Ext\left(B^0_{v,p} \cap \mathcal{C}^0\right), \forall v \in V, \forall p \in \mathscr{P} \right\}.$$

Theorem 2.5.9 *Let $(S_x)_{x \in X}$ be the family of all maximal $(\mathcal{M}, \mathcal{C})$-antialgebraic subsets of X. Then, for any $f \in CV_0(X, E)$ we have*

$$\overline{\mathcal{C}} = \left\{ f \in CV_0(X, E); f|S_x \in \overline{\mathcal{C}|S_x}, \forall x \in X \right\}.$$

Remark 2.5.10 If $\mathcal{M} \subset C(X, [0, 1])$ is a complemented family (i.e., $(1 - m) \in \mathcal{M}, \forall m \in \mathcal{M}$), and the convex cone $\mathcal{C} \subset CV_0(X, E)$ is stable with respect to the multiplication with elements of $\mathcal{M}$ or simply $\mathcal{M} \cdot \mathcal{C} \subset \mathcal{C}$, then for any $x \in X$ we have

$$S_x = [x]_M = \{ y \in X; m(y) = m(x), \forall m \in \mathcal{M} \}.$$

Corollary 2.5.11 *If $\mathcal{M} \subset C(X, [0, 1])$ is a complemented family, and the convex cone $\mathcal{C} \subset CV_0(X, E)$ is stable with respect to the multiplication with elements of $\mathcal{M}$, then we have*

$$\overline{\mathcal{C}} = \left\{ f \in CV_0(X, E); f \,\middle|\, [x]_M \in \overline{\mathcal{C}|\,[x]_{\mathcal{M}}}, \forall x \in X \right\}.$$

If $\mathcal{C}$ is a convex cone in $CV_0(X, E)$, for any $x \in X$ we denote by $\overline{\mathcal{C}(x)}$ the closure in E of the set:

$$\mathcal{C}(x) = \{ c(x); c \in \mathcal{C} \}.$$

Theorem 2.5.12 *If $\mathcal{M} \subset C(X, [0, 1])$ is a complemented family, and the convex cone $\mathcal{C} \subset CV_0(X, E)$ is such that $\mathcal{M} \cdot \mathcal{C} \subset \mathcal{C}$, then we have*

$$\overline{\mathcal{C}} = \left\{ f \in CV_0(X, E); \; f(x) \in \overline{\mathcal{C}(x)}, \forall x \in X \right\}.$$

Approximation of Continuously Differentiable Functions

3

3.1 Preliminaries and Notations

In this chapter, we study some approximation properties for different classes of differentiable real functions defined on open set Ω of $\mathbb{R}^n$, $n \in \mathbb{N}^*$.

For any element $k = (k_1, \ldots, k_n) \in \mathbb{N}^n$, called multi-index, we denote by $|k| = k_1 + \ldots + k_n$ and by D^k the derivative operator:

$$D^k = \frac{\partial^{|k|}}{\partial x_1^{k_1} \partial x_2^{k_2} \ldots \partial x_n^{k_n}}.$$

For any $p \in \mathbb{N}$, we denote by $C^p(\Omega)$ the set of all real functions on Ω which have partial continuous derivatives D^k for all $k = (k_1, \ldots, k_n)$, $|k| \le p$. We denote $C^\infty(\Omega) = \bigcap_{p \in N} C^p(\Omega)$ and we refer to a function $f \in C^\infty(\Omega)$ as being indefinitely derivable. Also, we denote by $C_0^p(\Omega)$ the set of all functions $f \in C^p(\Omega)$ with compact support, i.e., the set $[\,|f| > 0] = \{x \in \Omega;\ f(x) \ne 0\}$ is relatively compact, and its closure in $\mathbb{R}^n$ is included in Ω.

Obviously, $C_0^p(\Omega)$ may be identified with the set of all functions $f \in C_0^p(\mathbb{R}^n)$ such that $\overline{[\,|f| > 0]} \subset \Omega$. The functions $C_0^\infty(\mathbb{R}^n)$ are called test functions in $\mathbb{R}^n$.

The following function, introduced by Cauchy, is the most known indefinite differentiable function on R such that $f^{(p)}(0) = 0$, $\forall p \in \mathbb{N}$, which cannot be developed around the point $x_0 = 0$ in Taylor series:

$$f(x) = \begin{cases} e^{\frac{1}{x}} & \text{if } x < 0 \\ 0 & \text{if } x \ge 0. \end{cases}$$

I. Bucur, G. Paltineanu, *Topics in Uniform Approximation of Continuous Functions*, Frontiers in Mathematics, https://doi.org/10.1007/978-3-030-48412-5_3

Using this function, we may define the test function φ on $\mathbb{R}^n$:

$$\varphi(x) = \begin{cases} f\left(\|x\|^2 - 1\right) & \text{if} \ \ \|x\| < 1 \\ 0 & \text{if} \ \ \|x\| \geq 1, \end{cases}$$

where, for any $x = (x_1, x_2, \ldots, x_n) \in \mathbb{R}^n$, we have written $\|x\| = \sqrt{\sum_{i=1}^n x_i{}^2}$.

This function belongs to $C_0^\infty(\mathbb{R}^n)$, $\varphi \geq 0$ and its support is $\operatorname{supp}\varphi = \{x \in \mathbb{R}^n; \ \|x\| \leq 1\}$. Dividing the function φ by the positive number $\int_{\mathbb{R}} \varphi(x)dx$, we may suppose also that $\int \varphi dx = 1$.

For any $r > 0$, we denote by φ_r the function on $\mathbb{R}^n$ given by

$$\varphi_r(x) = \frac{1}{r^n} \cdot \varphi\left(\frac{x}{r}\right).$$

Obviously, this function has similar properties like φ but $\operatorname{supp}\varphi_r = \{x \in \mathbb{R}^n; \ \|x\| \leq r\}$.

If for any set $S \subset \mathbb{R}^n$ and any positive number r we denote by $B_r(S)$ the subset of $\mathbb{R}^n$ given by

$$B_r(S) = \left\{ y \in \mathbb{R}^n; \ d(y, S) < r \right\},$$

then $B_r(S)$ is an open subset of $\mathbb{R}^n$, and we have

$$\overline{B_r(S)} = \left\{ y \in \mathbb{R}^n; \ d(y, S) \leq r \right\}.$$

If S is bounded, then $B_r(S)$ is bounded, and the diameter $\delta\ [B_r(S)]$ of the set $B_r(S)$ is dominated by the number $2 \cdot r + \delta(S)$.

For any compact set $S \subset \mathbb{R}^n$ and any open set $\Omega \subset \mathbb{R}^n$ such that $S \subset \Omega$, we have

$$\overline{B_r(S)} \subset \Omega,$$

for any $r > 0$ such that

$$r < d\left(S, \mathbb{R}^n \backslash \Omega\right) = \inf\left\{d(s, y); \ s \in S, \ y \in \mathbb{R}^n \backslash \Omega\right\}.$$

Since the function $h : \mathbb{R}^n \to \mathbb{R}$, given by $h(x) = d(x, \mathbb{R}^n \backslash \Omega)$ is continuous, and the set $\mathbb{R}^n \backslash \Omega$ is closed, we have $h(x) > 0$, $\forall x \in \Omega$, and therefore the minimum of this function on S is strictly positive, i.e., $d(S, \mathbb{R}^n \backslash \Omega) > 0$.

Theorem 3.1.1 *For any compact K and any open subset Ω of $\mathbb{R}^n$, $K \subset \Omega$, there exists a function $\psi \in C_0^\infty(\Omega)$, $0 \leq \psi \leq 1$ and $\psi = 1$ on K.*

Proof Let us take $r \in \mathbb{R}$, $r > 0$ such that $3 \cdot r < d(K, \mathbb{R}^n \backslash \Omega)$. From Uryson's Lemma, there exists a continuous function $u : \mathbb{R}^n \to [0, 1]$ such that $u = 1$ on $B_r(K)$ and $u = 0$ on $B_{2 \cdot r}(K)$. Using the above function φ_r, we consider the function $U_r : \mathbb{R}^n \to \mathbb{R}$ given by

$$U_r(x) = (\varphi_r * u)(x) = \int \varphi_r(x - y) \cdot u(y) dy = \int \varphi_r(y) \cdot u(x - y) dy.$$

Since $\varphi_r \in C_0^\infty(\mathbb{R}^n)$, we deduce that $U_r \in C^\infty(\mathbb{R}^n)$. Moreover, we have $0 \leq U_r(x) \leq 1$ since $0 \leq u(x) \leq 1$ and $\int \varphi_r(y) dy = 1$.

If we take $x \in K$, $y \in B_r(0)$, we have $x - y \in B_r(K)$ and therefore, taking in account that $u(x - y) = 1$ if $y \in B_r(0)$, then

$$U_r(x) = \int_{B_r(x)} \varphi_r(x - y) \cdot u(y) dy = \int_{B_r(0)} \varphi_r(y) \cdot u(x - y) dy = \int_{B_r(0)} \varphi_r(y) dy = 1.$$

Analogously, if $x \notin B_{3 \cdot r}(K)$ and $y \in B_r(0)$, we have $x - y \notin B_{2 \cdot r}(K)$, $u(x - y) = 0$, and therefore $U_r(x) = 0$. The function U_r satisfied the required conditions since $U_r = 0$ outside the compact set $\overline{B_{3 \cdot r}(K)}$ included in Ω. $\qquad\square$

Theorem 3.1.2 *Let $\Omega \subset \mathbb{R}^n$ be an open set and let $K \subset \Omega$ be a compact subset. Then, for any function $f \in C^m(\Omega)$ there exists a function $F \in C^m(\mathbb{R}^n)$ such that $D^p f$ and $D^p F$ coincide on K for all $p = (p_1, \ldots, p_n) \in \mathbb{N}^n$, with $0 \leq |p| \leq m$. More precisely, we may choose $F \in C_0^m(\Omega)$.*

Proof Let $r \in \mathbb{R}$, $r > 0$ be such that $3 \cdot r < d(K, \mathbb{R}^n \backslash \Omega)$. Since $\overline{B_r(K)}$, is a compact subset of the open set $B_{2 \cdot r}(K)$ we may consider, using Theorem 3.1.1, a function $U \in C_0^\infty(B_{2 \cdot r}(K))$ such that $U = 1$ on $B_r(K)$. Since $B_{3 \cdot r}(K) \subset \Omega$, since the function $f \cdot U \in C^m(\Omega)$ and $f \cdot U = 0$ on $\Omega \backslash \overline{B_{2 \cdot r}(K)}$, we deduce that the function F on $\mathbb{R}^n$, equal to $f \cdot U$ on Ω and equal to 0 outside Ω, satisfies the required conditions.

The following assertion regards the decomposition of the unit function as a sum of functions from $C_0^\infty(\mathbb{R}^n)$ subordinated to a covering with open sets. $\qquad\square$

Theorem 3.1.3 *Let $K \subset \mathbb{R}^n$ be a compact set and let $\Omega_1, \ldots, \Omega_m$ a finite system of open sets of $\mathbb{R}^n$ such that $K \subset \bigcup_{I=1}^m \Omega_i$. Then, there exists $\varphi_i \in C_0^\infty(\Omega_i)$, $i = \overline{1, m}$ such that $\varphi_i \geq 0$, $\sum_{i=1}^m \varphi_i \leq 1$ on $\mathbb{R}^n$ and $\sum_{i=1}^m \varphi_i = 1$ on K.*

Proof For any $x \in K$ we denote by $V(x)$ a relatively compact neighbourhood of x such that $\overline{V(x)}$ is included in any open set Ω_i which contains x. Since $K \subset \bigcup_{x \in K} V(x)$ and K is compact, we may select a finite number $V(x_1), \ldots, V(x_r)$ such that for any i we denote $U_i = \bigcup \{V(x_j); \ V(x_j) \subset \Omega_i\}$. Obviously, $\overline{U_i}$ is a compact subset of Ω_i. From Theorem 3.1.1, there exists $\psi_i \in C_0^\infty(\Omega_i)$ such that $0 \leq \psi_i \leq 1$ and $\psi_i = 1$ on $\overline{U_i}$.

Furthermore, we denote

$$\varphi_1 = \psi_1, \ \varphi_i = \psi_i \cdot (1 - \psi_1) \cdot \ldots \cdot (1 - \psi_{i-1}), \ \forall i = \overline{2, m}.$$

Obviously, $\varphi_i \geq 0, \ \varphi_i \in C_0^\infty(\Omega_i), \ i = \overline{1, m}$, and

$$\sum_{i=1}^{m} \varphi_i = 1 - (1 - \psi_1) \cdot (1 - \psi_2) \cdot \ldots \cdot (1 - \psi_m).$$

We get $\sum_{i=1}^{m} \varphi_i \leq 1$ on $\mathbb{R}^n$ and $\sum_{i=1}^{m} \varphi_i = 1$ on K. $\qquad\qquad\square$

3.2   Bernstein Theorem

Let us denote by $C^m(\mathbb{R}^n)$ the algebra of all real functions on $\mathbb{R}^n$ possessing continuous derivatives D^p for any $p \in \mathbb{N}^n$ with $|p| \leq m$. For any compact $K \subset \mathbb{R}^n$ and any $f \in C^m(\mathbb{R}^n)$, we denote

$$\|f\|_K = \sup\left\{\left|D^p f(y)\right|; \ y \in K, \ |p| \leq m\right\}.$$

Obviously, the map $f \rightarrow \|f\|_K : C^m(\mathbb{R}^n) \rightarrow \mathbb{R}_+$ is a seminorm on $C^m(\mathbb{R}^n)$.

Definition 3.2.1 The compact convergence topology of order m, $m \in \mathbb{N}$ on $C^m(\mathbb{R}^n)$ is the locally convex topology generated by the seminorms $\{\| \|_K ; \ K \in \mathscr{K}\}$ where $\mathscr{K}$ is the set of all compact subsets of $\mathbb{R}^n$.

Since the countable family of seminorms $\{p_k\}_{k \in \mathbb{N}^*}$, $p_k = \| \|_{\overline{B(0,k)}}$ where

$$\overline{B(0, k)} = \left\{x \in \mathbb{R}^n; \ \sum_{i=1}^{n} x_i^2 \leq k^2\right\},$$

generates the above topology, one can see that the linear space $C^m(\mathbb{R}^n)$ endowed with the above topology is a Fréchet space, i.e. any Cauchy sequence of $C^m(\mathbb{R}^n)$ is convergent.

Obviously, the polynomial real functions on $\mathbb{R}^n$ belong to $C^m(\mathbb{R}^n)$ for any $m \in \mathbb{N}$, but we do not expect to approximate uniformly on $\mathbb{R}^n$ any element of $C^m(\mathbb{R}^n)$ with polynomial functions.

Nevertheless, such kind of approximation may be done with respect to the above compact topology on $C^m(\mathbb{R}^n)$.

Theorem 3.2.2 (Bernstein, see e.g. [26, p. 104]) *The algebra $\mathscr{P}(\mathbb{R}^n)$ of real polynomials is dense in $C^m(\mathbb{R}^n)$ with respect to the compact convergence topology of order m on $C^m(\mathbb{R}^n)$.*

Proof Let $f \in C^m(\mathbb{R}^n)$ and let K be a compact subset of $\mathbb{R}^n$. Using Theorem 3.1.2, we may consider a function $g \in C_0^m(\mathbb{R}^n)$ such that $g = f$ on a neighbourhood of K. Let K_1 be the support of g and let $d(K_1)$ be the diameter of K_1. With no loss of generality, we may suppose that $d(K_1) < 1$. We denote by I_1 the interval of $\mathbb{R}^n$ given by

$$I_1 = [-1, 1]^n = [-1, 1] \times [-1, 1] \times \ldots \times [-1, 1] \subset \mathbb{R}^n.$$

Also, for any $\varepsilon \in \mathbb{R}$, $0 < \varepsilon < 1$, we denote $I_\varepsilon = [-\varepsilon, \varepsilon]^n$. For any natural number n, we define the polynomial function $p_r : \mathbb{R}^n \to \mathbb{R}$:

$$p_r(x_1, \ldots, x_n) = c_r \cdot (1 - x_1{}^2)^r \cdot (1 - x_2{}^2)^r \cdot \ldots \cdot (1 - x_n{}^2)^r,$$

where c_r is chosen such that $\int_{I_1} p_r(x) dx = 1$, namely

$$c_r \cdot \left(\int_{-1}^{1} (1 - t^2)^r dt \right)^n = 1, \quad c_r = \left(\int_{-1}^{1} (1 - t^2)^r dt \right)^{-n}.$$

Obviously, for any $\delta \in (0, 1)$ we have

$$c_r < \left(\int_{-\delta}^{\delta} (1 - t^2)^r dt \right)^{-n} < \left(2 \cdot \delta \cdot (1 - \delta^2)^r \right)^{-n} = \frac{1}{(2 \cdot \delta)^n \cdot (1 - \delta^2)^{n \cdot r}}.$$

If $x = (x_1, \ldots, x_n) \in I_1 \backslash I_\varepsilon$, then at least for one $i = \overline{1, n}$, we have $x_i > \varepsilon$ and therefore

$$1 - x_i{}^2 < 1 - \varepsilon^2, \quad p_r(x) \le c_r \cdot (1 - \varepsilon^2)^r < \left(\frac{1 - \varepsilon^2}{(1 - \delta^2)^n} \right)^r \cdot \frac{1}{(2 \cdot \delta)^n}.$$

If ε is given $0 < \varepsilon < 1$, we can choose $\delta \in (0, 1)$ such that $\frac{1 - \varepsilon^2}{(1 - \delta^2)^n} < 1$. Since $\lim_{r \to \infty} \left(\frac{1 - \varepsilon^2}{(1 - \delta^2)^n} \right)^r = 0$, we deduce

$$\lim_{r \to \infty} \sup_{x \in I_1 \backslash I_\varepsilon} p_r(x) \le \lim_{r \to \infty} \frac{1}{(2 \cdot \delta)^n} \cdot \left(\frac{1 - \varepsilon^2}{(1 - \delta^2)^n} \right)^2 = 0.$$

Furthermore, for any $r \in \mathbb{N}^*$ we denote by $Q_r g$ the polynomial on $\mathbb{R}^n$:

$$(Q_r g)(x) = \int_{\mathbb{R}^n} p_r(x - y) \cdot g(y) dy = \int_{\mathbb{R}^n} p_r(z) \cdot g(x - z) dz, \quad \forall x \in \mathbb{R}^n.$$

We remark that for any multi-index p, $|p| \le m$ we have

$$D^p(Q_r g)(x) = \int p_r(z) \cdot (D^p g)(x-z)dz = [Q_r(D^p g)](x), \, x \in \mathbb{R}^n.$$

If $x \in K_1$ and $g(x-y) \ne 0$, then $x-y \in K_1$ and therefore

$$\|y\| = \|x - (x-y)\| \le d(K_1) < 1, \text{ i.e., } y \in I_1.$$

Hence for any $x \in K_1$ we have

$$
\begin{aligned}
|(Q_r g)(x) - g(x)| &= \left| \int_{I_1} p_r(z) \cdot g(x-z)dz - \int_{I_1} p_r(z) \cdot g(x)dz \right| \\
&= \left| \int_{I_1} p_r(z) \cdot [g(x-z) - g(x)]dz \right| \\
&\le \int_{I_1 \setminus I_\varepsilon} p_r(z) \cdot |g(x-z) - g(x)| \, dz \\
&\quad + \int_{I_\varepsilon} p_r(z) \cdot |g(x-z) - g(x)| \, dz \le 2 \cdot \|g\| \cdot \sup_{z \in I_1 \setminus I_\varepsilon} p_r(z) \\
&\quad + \sup_{z \in I_\varepsilon} |g((x-z) - g(x)| \\
&\quad \cdot \int_I p_r(z)dz.
\end{aligned}
$$

Similarly, for any multi-index p, $|p| \le m$, we have

$$|D^p Q_r g(x) - D^p g(x)| \le 2 \cdot \|D^p g\| \cdot \sup_{z \in I_1 \setminus I_\varepsilon} p_r(z) + \sup_{z \in I_\varepsilon} |D^p g(x-z) - D^p g(x)|.$$

The proof is finished using the above steps/conclusions, taking into account that the functions $D^p g$ are uniformly bounded and uniformly continuous on $\mathbb{R}^n$, and g coincides with f on a neighbourhood of K. □

Definition 3.2.3 On the space $C^\infty(\mathbb{R}^n)$, we also introduce a topology called compact convergence topology whose family of seminorms $\left(p_K^m\right)_{m \in \mathbb{N}, K \in \mathcal{K}}$ is given by

$$p_K^m(f) = \sup\left\{\left|D^p f(x)\right|, \, x \in K, \, p \text{ multi-index}, \, |p| \le m\right\}, \, \forall f \in C^\infty(\mathbb{R}^n).$$

Obviously, the sequence $\left(p_{\overline{B(0,m)}}^m\right)_{m \in \mathbb{N}}$ gives the same topology on $C^\infty(\mathbb{R}^n)$, and the locally convex space $C^\infty(\mathbb{R}^n)$ endowed with the above topology is a Fréchet space.

Theorem 3.2.4 *The algebra* $\mathscr{P}(\mathbb{R}^n)$ *of all real polynomials is dense in* $C^\infty(\mathbb{R}^n)$ *with respect to the compact convergence topology.*

Proof We consider a function $f \in C^\infty(\mathbb{R}^n)$, and for any natural number $m \in \mathbb{N}^*$ we choose, using Theorem 3.2.2, a polynomial P_m such that

$$p_{\overline{B(0,m)}}^m (P_m - f) < \frac{1}{m}.$$

Using the fact that the seminorms $\left(p_{\overline{B(0,m)}}^m \right)_{m \in \mathbb{N}}$ are a fundamental system of continuous seminorms on $C^\infty(\mathbb{R}^n)$, we conclude that

$$\lim_{m \to \infty} P_m = f. \qquad \qquad \Box$$

3.3   Stone-Weierstrass Theorem for Continuously Differentiable Functions

For any $m \in \mathbb{N}^*$ and any $f \in C^m(\mathbb{R}^n)$, we denote by $df(x)$ the first order differential of f at the point $x \in \mathbb{R}^n$. So $df(x) : \mathbb{R}^n \to \mathbb{R}$ is the following linear functional:

$$[df(x)](y) = \frac{\partial f}{\partial x_1}(x) \cdot y_1 + \ldots + \frac{\partial f}{\partial x_n}(x) \cdot y_n, \ \forall y = (y_1, \ldots, y_n) \in \mathbb{R}^n.$$

Theorem 3.3.1 (Nachbin, see e.g. [26, p. 107]) *Any subalgebra* $\mathcal{A} \subset C^m(\mathbb{R}^n)$ *with the properties:*

(i) $\forall x \in \mathbb{R}^n, \exists a_x \in \mathcal{A}$ *such that* $a_x(x) \neq 0$
(ii) $\mathcal{A}$ *separates the points of* $\mathbb{R}^n$,
(iii) $\forall x, y \in \mathbb{R}^n, y \neq 0, \exists a \in \mathcal{A}$ *such that* $[da(x)](y) \neq 0$ *is dense in* $C^m(\mathbb{R}^n)$ *with respect to the compact convergence topology of order* m.

Proof Let $K \subset \mathbb{R}^n$ be a compact subset, let U be an open, connected, and relatively compact neighbourhood of K, i.e., $U \supset K$. For instance U may be the set $B(0, r)$ for $r > 0$ sufficiently large. Using the compactness of $\overline{U}$, the property (i) and the continuity of any element of $\mathcal{A}$, we may consider a finite system $(f_1, f_2, \ldots, f_n)$ of elements of $\mathcal{A}$ such that

$$(f_1(x), \ldots, f_p(x)) \neq (0, \ldots, 0), \ \forall x \in \overline{U}.$$

Furthermore, we use the following assertion:

If $l_1, l_2, \ldots, l_k, \ k \leq n$ are linear functionals on $\mathbb{R}^n$, then they are linearly independent iff the dimension of linear subspace $\bigcap_{i=1}^k \text{Ker}(l_i)$ is $n - k$.

Let now $x_0 \in \mathbb{R}^n$ be a fixed element, and let $y_1 \in \mathbb{R}^n$ and $b_1 \in \mathcal{A}$ such that

$$y_1 \neq 0 \text{ and } db_1(x_0)(y_1) \neq 0.$$

If $n > 1$, we have $\dim \operatorname{Ker}[db_1(x_0)] = n - 1$ and we may consider $y_2 \in \operatorname{Ker}[db_1(x_0)]$, $y_2 \neq 0$, and using (ii) we may choose $b_2 \in \mathcal{A}$ such that $db_2(x_0)(y_2) \neq 0$. If $n > 2$, and since $db_2(x_0)$ and $db_1(x_0)$ are obviously independent, we get $\dim(\operatorname{Ker}[db_1(x_0)] \cap \operatorname{Ker}[db_2(x_0)]) = n - 2$, and therefore we may consider an element $y_3 \neq 0$ such that $db_1(x_0)(y_3) = 0$ and $db_2(x_0)(y_3) = 0$.

Using again (ii) we may choose $b_3 \in \mathcal{A}$ such that $db_3(x_0)(y_3) \neq 0$. The last relation shows that the linear functional $db_3(x_0)$ is not a linear combination of $db_1(x_0)$ and $db_2(x_0)$, i.e., $db_1(x_0)$, $db_2(x_0)$, $db_3(x_0)$ are linearly independent. We may continue this procedure and we construct $b_1, \ldots, b_n \in \mathcal{A}$ and $y_1, \ldots, y_n \in \mathbb{R}^n \setminus \{0\}$ such that

$$y_{i+1} \in \operatorname{Ker}[db_j(x_0)], \quad j = \overline{1, n}, \text{ and } db_{i+1}(x_0)(y_{i+1}) \neq 0.$$

The linear functional $db_{i+1}(x_0)$ is linearly independent of $db_1(x_0), db_2(x_0), \ldots, db_i(x_0)$, and therefore the system $db_1(x_0), db_2(x_0), \ldots, db_i(x_0), db_{i+1}(x_0)$ is linear independent for any $i \leq n - 1$, i.e., $(db_i(x_0))_{1 \leq i \leq n}$ is linearly independent.

Let now $x_0 \in \overline{U}$ and let $F : \mathbb{R}^n \to \mathbb{R}^n$ the vector function:

$$F(x) = (b_1(x), b_2(x), \ldots, b_n(x)), \quad \forall x \in \mathbb{R}^n.$$

If we denote by $y_0 = F(x_0)$, then from the local inversion theorem, we deduce the existence of two open subsets V_0 and W_0 such that $x_0 \in V_0$ and $y_0 \in W_0$ with the property that $F : V_0 \to W_0$ is a C^m-diffeomorphism.

Using the compactness of $\overline{U}$, we deduce the existence of a finite number $r \in \mathbb{N}^*$ of open subsets $V_1, V_2, \ldots, V_r$ and $W_1, W_2, \ldots, W_r$ such that $\overline{U} \subset \bigcup_{i=1}^r V_i$ and a C^m-diffeomorphism F_i between V_i and W_i, for any $i = \overline{1, r}$.

Furthermore, for any $i = \overline{1, r}$, we denote by $\left(b_j^i\right)_{1 \leq j \leq n}$ the scalar components of the vector function F_i, i.e.

$$F_i = (b_1^i, b_2^i, \ldots, b_n^i),$$

and we also denote

$$f_{p+(i-1) \cdot n + j} = b_j^i, \ 1 \leq i \leq r, \ 1 \leq j \leq n.$$

We consider now the compact set $\Omega = \overline{U} \times \overline{U} \setminus \bigcup\limits_{i=1}^{r} (V_i \times V_i)$. If $(x, y) \in \Omega$, then $x \neq y$ since we have $\overline{U} \subset \bigcup_{i=1}^{r} V_i$, and therefore there exists $g \in \mathcal{A}$ such that $g(x) \neq g(y)$. The elements of $\mathcal{A}$ being continuous functions, there exists an open neighbourhood $\tilde{V}$ of the point (x, y) such that $g(s) \neq g(t)$, $\forall (s, t) \in \tilde{V}$. Since Ω is compact, we may select a finite number of open sets $\tilde{V}_1, \tilde{V}_2, \ldots, \tilde{V}_q$ such that $\Omega \subset \bigcup\limits_{i=1}^{q} \tilde{V}_i$ and the same number of functions $g_1, \ldots, g_q \in \mathcal{A}$ such that

$$(g_1(s), \ldots, g_q(s)) \neq (g_1(t), \ldots, g_q(t)), \forall (s, t) \in \Omega.$$

We add to the above system $f_{p+(i-1)\cdot n+j} = b_j{}^i$, $1 \leq i \leq r$, $1 \leq j \leq n$ the functions:

$$f_{p+n\cdot r+k} = g_k, k = \overline{1, q},$$

and we denote $N = p + n \cdot r + q$.

From the above considerations, we deduce that the vector function $\Phi : \mathbb{R}^n \to \mathbb{R}^N$ given by

$$\Phi(x) = (f_1(x), \ldots, f_N(x)), \forall x \in \mathbb{R}^n,$$

is an injection from U onto $\Phi(U)$, the origin of $\mathbb{R}^n$ does not belong to $\Phi(U)$, $f_i \in C^m(\mathbb{R}^n)$, and the range of the matrix $\left(\frac{\partial f_i}{\partial x_j} \right)_{i \leq N, j \leq n}$ is equal to n at any point $x \in U$.

We consider now the open set $\tilde{U}$ in $\mathbb{R}^N$ given by $\tilde{U} = U \times B_1$ where

$$B_1 = \left\{ (x_{n+1}, x_{n+2}, \ldots, x_N); \ |x_i| < 1, \ \forall i = \overline{n + 1, N} \right\}$$

and the map $\tilde{\Phi} : \tilde{U} \to \mathbb{R}^N$ given by

$$\tilde{\Phi}(x_1, \ldots, x_n, x_{n+1}, \ldots, x_N) = (f_1(x), f_2(x), \ldots, f_n(x), f_{n+1}(x)$$
$$+ x_{n+1}, f_{n+2}(x) + x_{n+2}, \ldots, f_N(x) + x_N),$$

where $x = (x_1, x_2, \ldots, x_n) \in U$. Obviously, the components of $\tilde{\Phi}$ are of C^m-class.

One can easily verify that the range of $\tilde{\Phi}$ is equal to N at any point from $\tilde{U} = U \times B_1$ and $\tilde{\Phi}$ is injective. Hence the image $D = \tilde{\Phi}\left(\tilde{U} \right)$ is an open subset of $\mathbb{R}^n$, which does not contain the origin of $\mathbb{R}^N$, and $\tilde{\Phi}$ is a C^m-diffeomorphism between $\tilde{U}$ and D. If $f \in C^m(\mathbb{R}^n)$, we shall denote also by f the function defined on $\tilde{U}$ by

$$f(x, x') = f(x), \forall x \in U, \forall x' \in B_1.$$

The function $h : D \to \mathbb{R}$ given by $h = f \circ \tilde{\Phi}^{-1}$ belongs to the class, and applying Bernstein theorem, there exists a sequence $(Q_k)_k$ of polynomials on $C^m(\mathbb{R}^N)$ such that the function h and the functions $D^p h$, $|p| \le m$ are uniformly approximated on the compact subset $\Phi(K)$ of D by the sequences $(Q_k)_k$, in particular $(D^p Q_k)_k$. Since the origin of $\mathbb{R}^N$ is not contained in D, we may suppose that the polynomials Q_k take the value 0 in this origin and therefore the functions $a_k = Q_k(f_1, f_2, \ldots, f_n)$ belong to $\mathcal{A}$, and so the sequence $(a_k)_k$ approximates the function f on the compact K set in the compact convergence topology of order m.   □

Approximation Theorems in Locally Convex Lattices

4

4.1 Real Locally Convex Lattices: Preliminaries and Notations

We recall that an order relation "$\leq$" on a real vector E is called compatible with the linear structure of the space if:

$$(i)\,\forall\ x, y \in E, \ \ x \leq y \implies x + z \leq y + z, \forall\, z \in E,$$

$$(ii)\ \forall\, x, y \in E, \ x \leq y, \implies \ \alpha \cdot x \leq \alpha \cdot y \ , \forall \alpha \in \mathrm{R}_+.$$

An ordered vector space is a real vector space endowed with an order relation compatible with the linear structure of the space. An ordered set E is called lattice if for any $x, y \in E$ there exists the upper bound $x \vee y \in E$ and the lower bound $x \wedge y \in E$. A vector lattice is any ordered vector space that is a lattice. Let E be a vector lattice.

For any $x \in E$, the elements $x_+ = x \vee 0, x_- = x \wedge 0, \mid x \mid = x \vee (-x)$ are called the positive part, the negative part and the modulus of x, respectively. For any $x, y \in E$, we have

$$x + y = x \vee y + x \wedge y.$$

I. Bucur, G. Paltineanu, *Topics in Uniform Approximation of Continuous Functions*,
Frontiers in Mathematics, https://doi.org/10.1007/978-3-030-48412-5_4

Furthermore, the main properties of the vector lattices are presented:

$$1) x = x_+ - x_-,$$
$$2) x_+ \wedge x_- = 0,$$
$$3)\ |x| = x_+ + x_-,$$
$$4)\ |x| = x_+ \vee x_-,$$
$$5)\ |x| = |-x|,$$
$$6)\ |x + y| \le |x| + |y|,$$
$$7)\ |\alpha \cdot x| = |\alpha| \cdot |x|,\ \forall \alpha \in \mathbb{R}$$
$$8)\ ||x| - |y|| \le |x - y|$$
$$9)\ \ x \wedge (y \vee z) = (x \wedge y) \vee (x \wedge z)$$
$$10)\ x \vee (y \wedge z) = (x \vee y) \wedge (x \vee z).$$

A vector sublattice of a vector lattice E is any vector subspace $E_0 \subset E$ that is sublattice (i.e., $x \vee y \in E_0$ and $x \wedge y \in E_0$, $\forall x, y \in E_0$) The necessary and sufficient condition for a vector subspace E_0 of a vector lattice E to be a sublattice is

$$x_+ \in E_0,\ \forall x \in E_0.$$

A subset $A \subset E$ is called a solid set if $|x| < |y|$ and $y \in A \Rightarrow x \in A$.
     A vector subspace $I \subset E$ that is a solid set is called ideal, i.e.,

(i) I is a vector subspace,

(ii) $|x| < |y|$ and $y \in I \Rightarrow x \in I$.

Any ideal is a vector sublattice. If E is a vector lattice, then the quotient space E/I is also a vector lattice with respect to the order relation defined by the cone $\pi_I(E_+)$, i.e.,

$$\pi_I(x) \ge 0 \text{ iff } x \ge 0$$

A locally convex lattice is a vector lattice which is both a Hausdorff locally convex space and has a basis of the origin consisting of solid and convex neighbourhoods.
     A seminorm p on a vector lattice E is called solid iff $\forall x, y \in E$ such that $|x| \le |y|$ implies $p(x) \le p(y)$.
     Let E be a vector lattice and let τ be a Hausdorff locally convex topology on E. The necessary and sufficient condition for E to be a locally convex lattice is that τ is defined by a family of solid seminorms.
     A locally convex lattice is called of type (M) if its topology has a basis of the origin consisting of sublattices. In a locally convex lattice of type (M) there exists a basis of the origin consisting of sublattices which are solid and convex.

A seminorm p on a vector lattice E is called of type (M) iff

$$p(x \vee y) = \max\left(p(x), p(y)\right), \ \forall x, y \in E_+.$$

Let E be a vector lattice and let τ be a Hausdorff locally convex topology on E. The necessary and sufficient condition for E to be a locally convex lattice of type (M) is that τ is defined by a family of solid and of type (M) seminorms.

Typical examples of locally convex lattices of type (M) are the weighted spaces.

We remember that if X is a Hausdorff locally compact space and V is a Nachbin family on X, then the weighted space $CV_0(X, \mathbb{R})$ is the space of all functions $f \in C(X, \mathbb{R})$ such that $f \cdot v$ vanishes at infinity for any $v \in V$.

Obviously, $CV_0(X, \mathbb{R})$ is a vector lattice with respect to the pointwise order relation:

$$f \leq g \text{ iff } f(x) \leq g(x), \ \forall x \in X.$$

The weighted topology ω_V is determined by the seminorms $\{p_v\}_{v \in V}$, where

$$p_v(f) = \sup\{|f(x)| \cdot v(x); \ x \in X\}, \ \ \forall f \in CV_0(X, \mathbb{R}),$$

and has a basis of open neighbourhoods of the origin $(B_v)_{v \in V}$ of the form:

$$B_v = \{f \in CV_0(X, \mathbb{R}); \ p_v(f) < 1\}.$$

If we suppose in addition that for any $x \in X$ there exists a weight $v \in V$ such that $v(x) > 0$, then the weighted topology ω_V is Hausdorff.

Clearly, the sets (B_v) are convex, solid and sublattices. We also remark that the seminorms $(p_v)_{v \in V}$ are seminorms of type (M). In the particular case when $X = K$ is a Hausdorff compact space, then the space $C(X, \mathbb{R})$ is a normed vector lattice of type (M) with respect to the pointwise order relation and with the norm

$$\|f\| = \sup\{|f(x)|; \ x \in K\}$$

.

4.2   Ideals in Weighted Spaces

Let $S \subset X$ be a closed set and let $I_S = \{f \in CV_0(X); \ f|S = 0\}$. Obviously, I_S is an ideal of $CV_0(X)$. We also remark that I_S is closed with respect to the weighted topology. Indeed, let $x_0 \in S$ be an arbitrary fixed point, and let $v_0 \in V$ be a weight with the property $v_0(x_0) > 0$. If $g \in \overline{I_S}$, then for any $\varepsilon > 0$ and any $v \in V$ there exists $f \in I_S$ such that

$$p_v(g - f) = \sup\{\,|g(x) - f(x)| \cdot v(x); \forall x \in X\} < \varepsilon.$$

In the particular case $x = x_0$ and $v = v_0$, it results

$$|g(x_0) - f(x_0)| \cdot v_0(x_0) = |g(x_0)| \cdot v_0(x_0) < \varepsilon.$$

As $\varepsilon > 0$ is arbitrary, we deduce that $g(x_0) = 0$, so $g \in I_S$.

The next result shows that any closed ideal of $CV_0(X, \mathbb{R})$ has the preceding form.

Theorem 4.2.1 *Let* $I \subset CV_0(X, \mathbb{R})$ *be an arbitrary ideal. Then, there exists a closed subset* $S_I \subset X$ *such that*

$$\bar{I} = \{f \in CV_0(X, \mathbb{R}); \ f \,|S_I = 0\}.$$

Particularly, if I *is closed, then* $I = \{f \in CV_0(X, \mathbb{R}); \ f \,|S_I = 0\}$ *for some closed subset* $S_I \subset X$.

Proof Let $M = C(X, [0, 1])$ and for any $x \in X$, let $[x]$ be the subset of constancy for the functions from M. Obviously, $[x] = \{x\}$, $\forall x \in X$. The set I being an ideal, we get $M \cdot I \subset I$. Using now Corollary 2.3.9, we have

$$\bar{I} = \left\{f \in CV_0(X, \mathbb{R}); \ f \,\big|\, [x] \in \overline{I \,|\, [x]}, \ \forall x \in X \right\}.$$

If we denote $S_I = \{x \in X; \ h(x) = 0, \ \forall h \in I\}$, we remark that $\overline{I \,|\, [x]} = \{0\}$ and $\overline{I \,|\, [x]} = \mathbb{R}$ if $x \in S_I$, in particular $x \in X \backslash S_I$, and therefore

$$\bar{I} = \{f \in CV_0(X, \mathbb{R}); \ f(x) = 0, \ \forall x \in S_I\}. \qquad \square$$

Remark 4.2.2 If $\mathcal{I}$ denotes the set of all closed ideals of $CV_0(X, \mathbb{R})$ and $\mathcal{F}$ denotes the set of all closed subsets of X, then the map:

$$I \to S_I = \{x \in X; \ h(x) = 0, \ \forall h \in I\}$$

is a bijection between $\mathcal{I}$ and $\mathcal{F}$, and a decreasing one:

$$I' \subset I'' \ \Leftrightarrow \ S_{I''} \subset S_{I'}.$$

This allows us to generalize some results involving a different type of closed subset of X (antisymmetric, interpolating, antialgebraic sets) to the abstract case of closed ideals in a locally convex lattice.

4.3 Antisymmetric Ideals of a Real Locally Convex Lattice of (M)-Type

Let E be a real locally convex lattice of type (M). For any closed ideal $I \subset E$, we denote by $\pi_I : E \to E/I$ the canonical mapping and by $\pi_I' : (E/I)' \to E'$ the adjoint mapping of π_I. The center $Z(E)$ of E is the algebra of all order bounded endomorphisms on E, that is, those operators $U \in L(E, E)$ for which there exists $\lambda_U > 0$ such that $|U(x)| \le \lambda_U \cdot |x|$, $\forall x \in E$. The lowest number $\lambda > 0$ with the property $|U(x)| \le \lambda \cdot |x|$, $\forall x \in E$, will be noted by $\|U\|$. The center $Z(E)$ is a normed algebra with respect to the norm $U \to \|U\| : Z(E) \to \mathbb{R}_+$. The real part of the center is denoted by $ReZ(E)$ and is defined by

$$ReZ(E) = Z(E)_+ - Z(E)_+.$$

Definition 4.3.1 For any closed ideal I of E and any operator $U \in ReZ(E)$, we define $U_I : E/I \to E/I$ by

$$U_I[\pi_I(x)] = \pi_I[U(x)], \ \forall x \in E.$$

Obviously, the operator U_I is well-defined since $U(I) \subset I$, and U_I is positive iff U is positive. Furthermore, for any subset $A \subset ReZ(E)$, we note

$$A_I = \{U_I; U \in A\}.$$

Remark 4.3.2 If $A \subset ReZ(E)$, then $A_I \subset ReZ(E/I)$.

Indeed, since $\pi_I : E \to E/I$ is a lattice homeomorphism, for any $U \in A$ and any $x \in E$, we have

$$|U_I(\pi_I(x))| = |\pi_I(U(x))| = \pi_I(|U(x)|) \le \lambda_U \cdot \pi_I(|x|) = \lambda_U \cdot |\pi_I(x)|,$$

so $U_I \in Z(E/I)$. Hence, for $U \in ReZ(E)$ it results that $U_I \in ReZ(E/I)$.

Definition 4.3.3 Let I and J be two closed ideals of E such that $I \subset J$. Then, the following two mappings can be defined:

(i) $\pi_{IJ} : E/I \to E/J, \pi_{IJ}[\pi_I(x)] = \pi_J(x), \ \forall x \in E$

(ii) $M_{IJ} : ReZ(E/I) \to ReZ(E/J), M_{IJ}(U)[\pi_J(x)] = \pi_{IJ}[U(\pi_I(x))],$

$\forall U \in ReZ(E/I)$

The fact that the range of M_{IJ} is included in $ReZ(E/J)$ follows from the inequality:

$$|M_{IJ}(U)(\pi_J(x))| = |\pi_{IJ}[U(\pi_I(x))]| = \pi_{IJ}\,|[U(\pi_I(x))]| \leq \pi_{IJ}[\lambda_U \cdot |\pi_I(x)|]$$
$$= \lambda_U \cdot \pi_{IJ}[|\pi_I(x)|] = \lambda_U \cdot |\pi_{IJ}[\pi_I(x)| = \lambda_U \cdot |\pi_J(x)|\,, \forall x \in E\,.$$

According to Theorem 4.2.1, in the particular case of weighted spaces, any closed ideal $I \subset CV_0(X, \mathbb{R})$ has the form $I = I_S = \{f \in CV_0(X, \mathbb{R});\ f\,|S = 0\}$, where $S \subset X$ is a closed subset.

Now we observe that in this case:

$$\pi_{I_S}(f) = \{g \in CV_0(X, \mathbb{R});\ g\,|S = f\,|S\,\}\,.$$

The next proposition shows the form of the center of the lattice $E = CV_0(X, \mathbb{R})$.

Proposition 4.3.4 *If $E = CV_0(X, \mathbb{R})$, then $Z(E) = C_b(X, \mathbb{R})$ where $C_b(X, \mathbb{R})$ is the Banach space of all real, continuous and bounded functions on X.*

Proof If $U \in Z[CV_0(X, \mathbb{R})]$, then $U : CV_0(X, \mathbb{R}) \to CV_0(X, \mathbb{R})$ is a linear, continuous and order bounded operator, i.e. there exists $\lambda_U > 0$ such that

$$|U(f)| \leq \lambda_U \cdot |f|\,, \ \forall f \in CV_0(X, \mathbb{R}). \tag{4.1}$$

For any point $x_0 \in X$, we denote by δ_{x_0} the Dirac measure on $CV_0(X, \mathbb{R})$, i.e.,

$$\delta_{x_0}(f) = f(x_0), \ \forall f \in CV_0(X, \mathbb{R}),$$

and by L_{U,x_0} the following functional:

$$L_{U,x_0}(f) = U(f)(x_0), \ \forall f \in CV_0(X, \mathbb{R}).$$

We observe, taking into account (4.1), that

$$\delta_{x_0}(f) = f(x_0) = 0 \text{ implies } L_{U,x_0}(f) = 0, \ \forall f \in CV_0(X, \mathbb{R}).$$

Hence there exists a real number $a_U(x_0) \in \mathbb{R}$ such that

$$L_{U,x_0} = a_U(x_0) \cdot \delta_{x_0}.$$

Therefore, we have

$$U(f)(x_0) = a_U(x_0) \cdot f(x_0), \ \forall f \in CV_0(X, \mathbb{R}).$$

Since the point $x_0 \in X$ is arbitrary, we infer that

$$U(f)(x) = a_U(x) \cdot f(x), \ \forall f \in CV_0(X, \mathbb{R}), \ \forall x \in X. \tag{4.2}$$

From (4.1) and (4.2), we deduce that $|a_U(x)| \le \lambda_U, \ \forall x \in X$, so $a_U \in C_b(X, \mathbb{R})$.

The mapping $U \rightarrow a_U : Z[CV_0(X, \mathbb{R})] \rightarrow C_b(X, \mathbb{R})$ is an isomorphism of Banach algebras, hence $Z[CV_0(X, \mathbb{R})] \simeq C_b(X, \mathbb{R})$.

Now we come back to the general case of the real locally convex lattices of type (M). Let A be a subset of $ReZ(E)$ containing 0 and let $F \subset E$ be a vector subspace.   □

Definition 4.3.5 A closed ideal $I \subset E$ is said to be antisymmetric with respect to the pair (A, F) (or (A, F)-antisymmetric) if, for any $U \in A$ with the property $U_I [\pi_I(F)] \subset \pi_I(F)$, there exists $\alpha \in \mathbb{R}$ such that

$$U_I = \alpha \cdot 1_{E/I}.$$

where $1_{E/I}$ denoted the identity operator on the quotient space E/I.

Clearly, the lattice E itself is an antisymmetric ideal with respect to any pair (A, F) for any subset $A \subset ReZ(E)$ containing 0 and any vector subspace $F \subset E$. Furthermore, we shall denote by $\mathcal{A}_{A,F}(E)$ the family of all closed ideals $I \subset E$, antisymmetric with respect to the pair (A, F).

Remark 4.3.6 Let $E = CV_0(X, \mathbb{R})$, $A \subset C_b(X, \mathbb{R})$ be a subset containing the constant function 0 and let $F \subset CV_0(X, \mathbb{R})$ be a vector subspace.

Then, a subset $S \subset X$ is (A, F)-antialgebraic in the sense of the Definition 2.3.23 iff the closed ideal $I_S = \{g \in CV_0(X, \mathbb{R}); g \,|S = 0\}$ is (A, F)-antisymmetric in the sense of the Definition 4.3.5.

Indeed, according to Proposition 4.3.4, the center $Z[CV_0(X, \mathbb{R})]$ is isomorphic with $C_b(X, \mathbb{R})$. More, from the proof of this Proposition, it results that the isomorphism $U \in Z[CV_0(X, \mathbb{R})] \leftrightarrow a_U \in C_b(X, \mathbb{R})$ consists in the equality:

$$U(f) = a_U \cdot f, \ \forall f \in CV_0(X, \mathbb{R}).$$

On the other hand, for any $h \in CV_0(X, \mathbb{R})$ we have $\pi_{I_S}(h) = h \,|S$, and further

$$U_{I_S}[\pi_{I_S}(f)] = \pi_{I_S}(U(f)) = \pi_{I_S}(a_U \cdot f) = (a_U \cdot f) \,|S .$$

Therefore

$$U_{I_S}[\pi_{I_S}(F)] \subset \pi_{I_S}(F) \Leftrightarrow a_U \cdot F \,|S \subset F \,|S .$$

We also remark that $U_{I_S} = \alpha \cdot 1_{CV_0(X,\mathbb{R})/I_s}$ is equivalent to the fact that $a_U(x) = \alpha, \ \forall x \in S$, so a_U is constant on S.

We return now to the general case of the locally convex lattices of type (M). Let A be a subset of $ReZ(E)$ containing 0 and let $F \subset E$ be a vector subspace.

Theorem 4.3.7 *Let $(I_\alpha)_\alpha$ be a family of elements of $\mathcal{A}_{A,F}(E)$ with the property $J = \sum_\alpha I_\alpha \neq E$.*
Then

$$I = \bigcap_\alpha I_\alpha \in \mathcal{A}_{A,F}(E).$$

Proof Let us consider the following commutative diagram:

$$ReZ(E/I) \xrightarrow{M_{II_\alpha}} ReZ(E/I_\alpha)$$

$$M_{IJ} \searrow \ \swarrow M_{I_\alpha J}$$

$$ReZ(E/J).$$

If $U \in A$ has the property $U_I \left[\pi_I(F)\right] \subset \pi_I(F)$, then

$$M_{II_\alpha}(U_I)\left[\pi_{I_\alpha}(F)\right] = \pi_{II_\alpha}\left[U_I(\pi_I(F))\right] \subset \pi_{II_\alpha}\left[\pi_I(F)\right] = \pi_{I_\alpha}(F).$$

For any $x \in E$ and any $U \in A$ we have

$$M_{II_\alpha}(U_I)[\pi_{I_\alpha}(x)] = \pi_{II_\alpha}[U_I(\pi_I(x))]$$

$$= \pi_{II_\alpha}[\pi_I(U(x))] = \pi_{I_\alpha}[U(x)] = U_{I_\alpha}[\pi_{I_\alpha}(x)].$$

Therefore,

$$M_{II_\alpha}(U_I) = U_{I_\alpha} \in A_{I_\alpha} \subset ReZ(E/I_\alpha) \text{ and } M_{II_\alpha}(U_I)\left[\pi_I(F)\right] \subset \pi_{I_\alpha}(F)$$

Since $I_\alpha \in \mathcal{A}_{A,F}(E)$, it follows that there exists a real number a_α such that

$$M_{II_\alpha}(U_I) = a_\alpha \cdot 1_{E/I_\alpha}.$$

Furthermore, we shall prove that

$$M_{IJ}(U_I) = a_\alpha \cdot 1_{E/J}.$$

Indeed, from the previous diagram, it results

$$M_{IJ}(U_I)[\pi_J(x)] = M_{I_\alpha J}[M_{II_\alpha}(U_I)][\pi_{I_\alpha}(x)]$$

$$= M_{I_\alpha J}(a_\alpha \cdot 1_{E/I_\alpha})[\pi_{I_\alpha}(x)] = a_\alpha \cdot M_{I_\alpha J}[\pi_{I_\alpha}(x)]$$

$$= a_\alpha \cdot \pi_J(x) = a_\alpha \cdot 1_{E/J}[\pi_J(x)], \ \forall x \in E$$

Since $J = \sum_\alpha I_\alpha \neq E$, we deduce that $E/J \neq \{0\}$ and furthermore, that $a_\alpha = a$ (constant) for any α. Therefore, we have

$$M_{II_\alpha}(U_I) = a \cdot 1_{E/I_\alpha} = a \cdot M_{II_\alpha}(1_{E/I}), \text{ so}$$

$$M_{II_\alpha}(U_I - a \cdot 1_{E/I}) = 0 \ \forall \alpha.$$

Since,

$$0 = M_{II_\alpha}(U - a \cdot 1_{E/I})[\pi_{I_\alpha}(x)] = \pi_{I_\alpha}[U(x) - a \cdot x], \forall \alpha,$$

it follows $U(x) - a \cdot x \in I_\alpha, \forall \alpha$, hence $U(x) - a \cdot x \in I, \forall x \in E$. Furthermore, it results that $\pi_I[U(x) - a \cdot x] = 0$, which is equivalent to $U_I[\pi_I(x)] - a \cdot 1_{E/I}[\pi_I(x)] = 0$, so

$$U_I = a \cdot 1_{E/I}.$$

Therefore, $I \in \mathcal{A}_{A,F}(E)$ and the proof is finished.   □

Corollary 4.3.8 *For any closed ideal* $I \in \mathcal{A}_{A,F}(E)$, $I \neq E$, *there exists a unique, minimal, closed ideal* $\tilde{I} \subset I$, *antisymmetric with respect to the pair* (A, F).

Proof Let $I \in \mathcal{A}_{A,F}(E)$ such that $I \neq E$. If we denote by $\tilde{I} = \cap \{J \in \mathcal{A}_{A,F}(E); \ J \subset I\}$, then, from Theorem 4.3.3, it follows that $\tilde{I} \in \mathcal{A}_{A,F}(E)$. On the other hand, obviously $\tilde{I} \subset I$ and $\tilde{I}$ is minimal.

Furthermore, we shall denote by $\tilde{\mathcal{A}}_{A,F}(E)$ the family of all minimal, closed ideals, (A, F)-antisymmetric.   □

The next theorem is a generalization of de Brange's lemma for locally convex lattices of type (M).

Theorem 4.3.9 *Let* $A \subset Re Z(E)$ *be a subset containing* 0, $F \subset E$ *a vector subspace, and let* W *be a convex and solid neighbourhood of the origin of* E *which also is a sublattice. If* $f \in Ext\{W^0 \cap F^0\}$ *and* $I = \{x \in E; \ |f| \ (|x|) = 0\}$, *then* $I \in \mathcal{A}_{A,F}(E)$.

Proof Let $U \in A$ be such that $U_I[\pi_I(F)] \subset \pi_I(F)$. Since $ReZ(E/I) = Z(E/I)_+ - Z(E/I)_+$ and $U_I \in ReZ(E/I)$, we can suppose that $0 \le U_I \le 1_{E/I}$.

Indeed, let $S, T \in Z(E/I)_+$ be such that $U_I = S - T$. Since S is an order bounded isomorphism, it follows that there exists $\lambda_S > 0$ with the property:

$$|S[\pi_I(x)]| \le \lambda_S \cdot |\pi_I(x)|, \forall x \in E.$$

Of course, we have

$$S[\pi_I(x)] \le \lambda_S \cdot \pi_I(x), \forall \pi_I(x) \ge 0.$$

Changing S by $\frac{1}{\lambda_S} \cdot S$, it results that $0 \le S \le 1_{E/I}$. On the other hand, it is clear that $S[\pi_I(F)] \subset \pi_I(F)$. Similarly, we can suppose that $0 \le T \le 1_{E/I}$ and $T[\pi_I(F)] \subset \pi_I(F)$. If we prove that $S = \alpha \cdot 1_{E/I}$ and $T = \beta \cdot 1_{E/I}$, then it results that $U_I = (\alpha - \beta) \cdot 1_{E/I}$ and so the proof of the theorem is finished.

Let $U_I \in Z(E/I)_+, 0 \le U_I \le 1_{E/I}$ such that $U_I[\pi_I(F)] \subset \pi_I(F)$. Since $f \in I^0 \subset E'$, it results that there exists $g \in (E/I)'$ such that $f = \pi_I'g$. Indeed, let $g : E/I \to \mathbb{R}$ given by

$$g[\pi_I(x)] = f(x), \forall x \in E.$$

Obviously, g is well-defined, since $\pi_I(x) = \pi_I(y)$, then $x - y \in I$, and taking into account that $f \in I^0$, it results that $f(x) = f(y)$.

Hence, there exists $g \in (E/I)'$ such that

$$f = g \circ \pi_I = \pi_I{}' \circ g.$$

Obviously, we have

$$g \in Ext\left\{[\pi_I(W)]^0 \cap [\pi_I(F)]^0\right\}.$$

Furthermore, we denote

$$g_1 = U_I'g, \text{ and } g_2 = (1_{E/I} - U_I)'g$$

and

$$a_i = \inf\left\{\lambda > 0; g_i \in \lambda \cdot [\pi_I(W)]^0\right\} = \sup\{|g_i[\pi_I(x)]|; x \in W\}, i = 1, 2.$$

Since $g = g_1 + g_2 \in (a_1 + a_2) \cdot [\pi_I(W)]^0$, it follows that $a_1 + a_2 \ge 1$.

On the other hand, for $x_1, x_2 \in W$ we have

$$|g_1[\pi_I(x_1)]| + |g_2[\pi_I(x_2)]| = |g[U_I(\pi_I(x_1))]| + |g[1_{E/I} - U_I](\pi_I(x_2))|$$

$$\leq |g|[U_I(|\pi_I(x_1)| \vee |\pi_I(x_2)|) + (1_{E/I} - U_I)(|\pi_I(x_1)| \vee |\pi_I(x_2)|)]$$

$$= |g|(|\pi_I(x_1)| \vee |\pi_I(x_2)|).$$

Since $\pi_I(W)$ is a solid subset and a sublattice, it follows that $|\pi_I(x_1)| \vee |\pi_I(x_2)| \in \pi_I(W)$. Taking into account that $g \in [\pi_I(W)]^0$, it results that $|g|(|\pi_I(x_1)| \vee |\pi_I(x_2)|) \leq 1$.

Therefore, we proved that

$$|g_1[\pi_I(x_1)]| + |g_2[\pi_I(x_2)]| \leq 1, \quad \forall x_1, x_2 \in W,$$

from which we deduce that $a_1 + a_2 \leq 1$ and so that $a_1 + a_2 = 1$.

Furthermore, we shall prove that $g_i \neq 0$, $i = 1, 2$. First, we shall prove that

$$\left|U_I^{'}(g)\right| = U_I^{'}(|g|).$$

Indeed, since $0 \leq U_I^{'} \leq 1_{E/I}$, it follows that $U_I^{'}(g_+) = (U_I^{'}g)_+$ and $U_I^{'}(g_-) = (U_I^{'}g)_-$. Furthermore, we have:

$$\left|U_I^{'}(g)\right| = U_I^{'}(g)_+ + U_I^{'}(g)_- = U_I^{'}(g_+) + U_I^{'}(g_-) = U_I^{'}(g_+ + g_-) = U_I^{'}(|g|).$$

If we suppose now that $g_1 = 0$, and taking into account that $U \geq 0$ (because $U_I \geq 0$), then for any $x \in E$ we have

$$|f|(|U(x)|) \leq |f|(U(|x|)) = |\pi_I^{'}g|(U(|x|)) = \pi_I^{'}|g|(U(|x|))$$

$$= |g|(U_I(\pi_I(|x|))) = |U_I^{'}g|(\pi_I(|x|)) = |g_1|(\pi_I(|x|)) = 0,$$

from which we deduce that $U(x) \in I$ and so $U_I = 0 = 1_{E/I}$. Therefore $I \in \mathcal{A}_{A,F}(E)$.

Similarly, if we suppose that $g_2 = 0$, then it results that $1_{E/I} - U_I = 0$ and so $I \in \mathcal{A}_{A,F}(E)$. Therefore, we can suppose further that $g_1 \neq 0$ and $g_2 \neq 0$, where from $a_i > 0$, $i = 1, 2$.

On the other hand, since $g \in [\pi_I(F)]^0$ and $U_I[\pi_I(F)] \subset \pi_I(F)$, it follows that

$$g_i \in [\pi_I(F)]^0, i = 1, 2.$$

Taking into account that $g = a_1 \cdot \frac{g_1}{a_1} + a_2 \cdot \frac{g_2}{a_2}$ and $g \in Ext\left\{[\pi_I(W)]^0 \cap [\pi_I(F)]^0\right\}$, we deduce that $g = \frac{g_1}{a_1}$ or $g = \frac{g_2}{a_2}$. But we observe that the equality $g_1 = a_1 \cdot g$ implies that $(a_1 \cdot 1_{E/I} - U_I)^{'} = 0$ and furthermore, that $U_I = a_1 \cdot 1_{E/I}$. Similarly, the equality

$g_2 = a_2 \cdot g$ implies that $U_I = a_2 \cdot 1_{E/I}$. Therefore, in both cases $I \in \mathcal{A}_{A,F}(E)$ and so the proof of the theorem is finished.   □

Corollary 4.3.10 *Let* $E = CV_0(X, \mathbb{R})$, $A \subset C_b(X, \mathbb{R})$ *be a subset containing* 0, *and* $F \subset E$ *be a vector subspace. If* $L = v \cdot \mu \in Ext\left\{B_v^{\,0} \cap F^0\right\}$, *then the support of* L *is an antialgebraic subset with respect to the pair* (A, F), *in the sense of Definition 2.3.23.*

Proof If we denote by

$$I = \{f \in CV_0(X, \mathbb{R}); \, |L|\,(|f|) = 0\},$$

then, according to the previous theorem, I is an ideal (A, F)-antisymmetric. On the other hand, if we denote by S_L the support of L, then we have

$$I = \{f \in CV_0(X, \mathbb{R}); \, f\,|_{S_L} = 0\}.$$

From the Remark 4.3.6, we deduce that the set S_L is an (A, F)-antialgebraic subset.   □

4.4   An Approximation Theorem for a Vector Subspace of a Real Locally Convex Lattice of (M)-Type

The main result of this section is the following Bishop's type theorem:

Theorem 4.4.1 *Let* E *be a real locally convex lattice of type* (M), $A \subset ReZ(E)$ *a subset containing* 0, *and let* $F \subset E$ *be a vector subspace. Then, for any* $x \in E$ *we have*

$$x \in \overline{F} \iff \pi_I(x) \in \overline{\pi_I(F)}, \; \forall I \in \widetilde{\mathcal{A}}_{A,F}(E),$$

where $\widetilde{\mathcal{A}}_{A,F}(E)$ *denoted the family of all minimal closed ideals* (A, F)-*antisymmetric.*

Proof The necessity of the assertion is obvious. In order to prove the sufficiency, we assume that $\pi_I(x) \in \overline{\pi_I(F)}, \forall I \in \widetilde{\mathcal{A}}_{A,F}(E)$ and $x \notin \overline{F}$. Then, there exists a $f \in E'$ such that $f(x) \neq 0$ and $f(y) = 0$, $\forall f \in F$. Since f is continuous, it follows that there exists a convex, solid neighbourhood W of the origin of E, which is a sublattice such that

$$|f|\,(|x|) \leq 1, \; \forall x \in W.$$

Since $f \in F^0$, it follows that $f \in W^0 \cap F^0$. Now, from Krein Milman theorem, it follows that we can suppose

$$f \in Ext\left\{W^0 \bigcap F^0\right\}.$$

If we denote by

$$J = \left\{ y \in E; \, |f| \, (|y|) = 0 \right\},$$

then, from Theorem 4.3.9, it results that J is an (A, F)-antisymmetric ideal.

On the other hand, from the Corollary 4.3.8, we deduce that there exists an ideal $J_0 \in \tilde{\mathcal{A}}_{(A,F)}(E)$ such that $J_0 \subset J$. Since $f \in J_0{}^0 \cap F^0$ and $\pi_{J_0}(x) \in \overline{\pi_{J_0}(F)}$ (by hypothesis), it follows that $f(x) = 0$.

Indeed, let $g : E/J_0 \to \mathbb{R}$ given by $g[\pi_{J_0}(x)] = f(x)$. We remark that f is well-defined, since $f \in J_0{}^0$. Also, we observe that $f \in F^0$ implies that $g \in [\pi_{J_0}(F)]^0$.

Taking into account that $\pi_{J_0}(x) \in \overline{\pi_{J_0}(F)}$, it results that $f(x) = g[\pi_{J_0}(x)] = 0$, but this is a contradiction, since $f(x) \neq 0$.

We remember that, according to Remark 4.3.6, for $E = CV_0(X, \mathbb{R})$, $A \subset C_b(X, \mathbb{R})$, with $0 \in A$ and $F \subset CV_0(X, \mathbb{R})$ a vector subspace, a closed subset $S \subset X$ is (A, F)-antialgebraic in the sense of Definition 2.3.23 iff the closed ideal $I_S = \{g \in CV_0(X, \mathbb{R}); g \,|S = 0\}$ is antisymmetric with respect to the pair (A, F) in the sense of Definition 4.3.5. Moreover, S is maximal iff I_S is minimal.   □

Corollary 4.4.2 *Let $A \subset C_b(X, \mathbb{R})$ be a subset containing 0 and let $F \subset CV_0(X, \mathbb{R})$ be a vector subspace. Then, for any $f \in CV_0(X, \mathbb{R})$ we have*

$$f \in \overline{F} \quad \Leftrightarrow \quad f \,|S_x \in F \,|S_x \,, \; \forall x \in X,$$

where $(S_x)_{x \in X}$ denoted the family of all maximal subsets of X antialgebraic with respect to the pair (A, F).

The assertion follows from Theorem 4.4.1 and the previous comment.

Furthermore, a Stone-Weierstrass type theorem for a locally convex lattice of type (M) is presented.

Theorem 4.4.3 *Let E be a real, locally convex lattice of (M)-type, $A \subset ReZ(E)$ a subset containing 0, and $F \subset E$ a vector subspace with the properties:*

(i) $AF = \{U(F); \, U \in A\} \subset F$,
(ii) *F is not included in any maximal ideal of E,*
(iii) *Any closed ideal $I \subset E$ such that $\{U_I; \, U \in A\} \subset \mathbb{R} \cdot 1_{E/I}$ is a maximal ideal. Then,*

$$\overline{F} = E.$$

Proof Let $x \in E$ be arbitrary and let $I \in \tilde{\mathcal{A}}_{A,F}(E)$. From hypothesis (i), we deduce that

$$U_I[\pi_I(F)] \subset \pi_I(F), \forall \, U \in A.$$

Since the closed ideal I is (A, F)-antisymmetric, it results that there exists a real number $a_U \in \mathbb{R}$ such that $U_I = a_U \cdot 1_{E/I}$ and so that I is a maximal ideal, according to (iii). On the other hand, it is known that if I is maximal, then the quotient space $E/I = \pi_I(E)$ has the dimension 1.

From hypothesis (ii), it follows that $\pi_I(F) \neq \{0\}$. Now, from the inclusions $\{0\} \subset \pi_I(F) \subset \pi_I(E)$ we deduce that $\pi_I(F) = \pi_I(E)$ and so $\pi_I(x) \in \pi_I(F)$.

Since $I \in \widetilde{\mathcal{A}}_{A,F}(E)$ was arbitrary, we deduce from Theorem 4.4.1 that $x \in \overline{F}$, and so the proof is finished.   □

In the particular case of weighted space, we obtain the following Stone-Weierstrass's type theorem.

Theorem 4.4.4 *Let A be a* subset *of $C_b(X, \mathbb{R})$ with the property $0 \in A$ and let F be a* vector subspace of $CV_0(X, \mathbb{R})$ such that

 (i) $AF = \{U(x); \forall U \in A, \forall x \in F\} \subset F$,
 (ii) A separates the points of X,
 (iii) For any $x \in X$ there exists a function $f \in F$ such that $f(x) \neq 0$. Then,

$$\overline{F} = CV_0(X, \mathbb{R}).$$

Proof Since the center of the lattice $E = CV_0(X, \mathbb{R})$ is $C_b(X, \mathbb{R})$ and $ReZ(E) = C_b(X, \mathbb{R})$, it results that $A \subset ReZ(E)$. On the other hand, from (iii) it follows that F is not included in any maximal ideal. Indeed, for any maximal ideal I of the weighted space $CV_0(X, \mathbb{R})$, there exists a point $x \in X$ such that $I = I_x = \{f \in CV_0(X, \mathbb{R}); f(x) = 0\}$. If we suppose that F is included in a maximal ideal, then there is a $x_0 \in X$ such that

$$F \subset I_{x_0} = \{f \in CV_0(X, \mathbb{R}); f(x_0) = 0\},$$

where from it results that $f(x_0) = 0$, $\forall f \in F$, which contradicts the hypothesis (iii). From (i) and (ii), it follows that any subset $S \subset X$, antisymmetric with respect to the pair (A, F), is a singleton, where from we deduce that the correspondent ideal I_S is maximal. The conclusion of the theorem follows now from Theorem 4.4.3.   □

4.5   An Approximation Theorem for a Convex Cone of a Real Locally Convex Lattice of (M)-Type

Definition 4.5.1 Let E be a real locally convex lattice of (M)-type, let A be a subset of $ReZ(E)$ containing 0 and $C \subset E$ a convex cone.

A closed ideal $I \subset E$ is called antisymmetric with respect to the pair $(A, \mathcal{C})$ if for any $U \in A$ with the properties:

$$(i)\, 0 \leq U_I \leq 1_{E/I},$$

$$(ii)\, U_I \left[\pi_I(\mathcal{C})\right] + \left[1_{E/I} - U_I\right]\left[\pi_I(\mathcal{C})\right] \subset \pi_I(\mathcal{C}),$$

there exists $\alpha \in \mathbb{R}$, such that

$$U_I = \alpha \cdot 1_{E/I}.$$

Furthermore, we shall denote by $\mathcal{A}_{A,\mathcal{C}}(E)$ the family of all closed ideals $I \subset E$, antisymmetric with respect to the pair $(A, \mathcal{C})$.

Theorem 4.5.2 *Let (I_α) be a family of elements of $\mathcal{A}_{A,\mathcal{C}}(E)$ such that $J = \sum_\alpha I_\alpha \neq E$. Then,*

$$I = \bigcap_\alpha I_\alpha \in \mathcal{A}_{A,\mathcal{C}}(E).$$

Proof Let $U \in A$ be with the properties:

$$(i)\, 0 \leq U_I \leq 1_{E/I},$$

$$(ii)\, U_I[\pi_I(\mathcal{C})] + [1_{E/I} - U_I][\pi_I(\mathcal{C})] \subset \pi_I(\mathcal{C}).$$

Furthermore, we have

$$M_{II_\alpha}(U_I)(\pi_{I_\alpha}(\mathcal{C})) + [1_{E/I_\alpha} - M_{II_\alpha}(U_I)](\pi_{I_\alpha}(\mathcal{C}))$$

$$= \pi_{II_\alpha}[U_I(\pi_I(\mathcal{C})] + \pi_{II_\alpha}[\pi_I(\mathcal{C})] - \pi_{II_\alpha}[U_I(\pi_I(\mathcal{C})]$$

$$= \pi_{II_\alpha}[U_I(\pi_I(\mathcal{C}) + (1_{E/I} - U_I)(\pi_I(\mathcal{C})] \subset \pi_{II_\alpha}[(\pi_I(\mathcal{C}))] = \pi_{I_\alpha}(\mathcal{C}).$$

Since I_α is an antisymmetric ideal with respect to $\mathcal{C}$, it follows that there exists $a_\alpha \in \mathbb{R}$ such that

$$M_{II_\alpha}(U_I) = a_\alpha \cdot 1_{E/I_\alpha}.$$

On the other hand, from the following commutative diagram:

$$ReZ(E/I) \xrightarrow{M_{II_\alpha}} ReZ(E/I_\alpha)$$

$$M_{IJ} \searrow \swarrow M_{I_\alpha J}$$

$$ReZ(E/J)$$

we have

$$M_{IJ}(U_I) = M_{I_\alpha J}(M_{II_\alpha}(U_I)) = M_{I_\alpha J}\left(a_\alpha \cdot 1_{E/I_\alpha}\right) = a_\alpha \cdot 1_{E/I_\alpha}.$$

Since $J = \sum_\alpha I_\alpha \neq E$, we deduce that $E/J \neq \{0\}$ and so that $a_\alpha = a$(constant), $\forall \alpha$. Hence we have

$$M_{II_\alpha}(U_I) = a \cdot 1_{E/I_\alpha} = a \cdot M_{II_\alpha}(1_{E/I}),$$

and further

$$M_{II_\alpha}(U_I - a \cdot 1_{E/I}) = 0 \forall \alpha.$$

Since $0 = M_{II_\alpha}\left(U_I - a \cdot 1_{E/I}\right)\left[\pi_{I_\alpha}(x)\right] = \pi_{I_\alpha}\left[U(x) - a \cdot x\right]$, $\forall \alpha$, we deduce that

$$U(x) - a \cdot x \in I_\alpha, \forall \alpha,$$

and so that $U(x) - a \cdot x \in I$, $\forall x \in E$. Therefore, $\pi[U(x) - a \cdot x] = 0$, which is equivalent to $U_I[\pi_I(x)] - a \cdot 1_{E/I}[\pi_I(x)] = 0$ and further with

$$U_I = a \cdot 1_{E/I}.$$

Therefore, $I \in \mathcal{A}_{A,\mathcal{C}}(E)$, and the proof is finished.   □

Remark 4.5.3 Let $E = CV_0(X, \mathbb{R}), A \subset C_b(X, \mathbb{R})$ be a subset containing 0 and let $\mathcal{C} \subset CV_0(X, \mathbb{R})$ be a convex cone with $0 \in \mathcal{C}$.

A subset $S \subset X$ is antialgebraic with respect to the pair $(A, \mathcal{C})$ in the sense of the Definition 2.3.23 iff the closed ideal $I_S = \{g \in CV_0(X, \mathbb{R}); \ g \,|S = 0\}$ is antisymmetric with respect to the pair $(A, \mathcal{C})$ in the sense of the Definition 4.5.1.

The proof is similar to the one in Remark 4.3.6. We return now to the general case of the locally convex lattices of type (M).

Corollary 4.5.4 *Every own closed ideal I, $(A, \mathcal{C})$-antisymmetric, contains a unique minimal ideal $\tilde{I}$, $(A, \mathcal{C})$-antisymmetric.*

Proof Let $I \in \mathcal{A}_{A,\mathcal{C}}(E)$ such that $I \neq E$. If we denote by $\tilde{I} = \bigcap\{J \in \mathcal{A}_{A,\mathcal{C}}(E); \ J \subset I\}$, then, according to Theorem 4.5.2, it follows that $\tilde{I} \in \mathcal{A}_{A,\mathcal{C}}(E)$. On the other hand, it is clear that $\tilde{I}$ is minimal and that $\tilde{I} \subset I$. Furthermore, we shall denote by $\tilde{\mathcal{A}}_{A,\mathcal{C}}(E)$ the family of all minimal, closed ideals of E, antisymmetric with respect to the pair $(A, \mathcal{C})$.   □

The next theorem is a generalization of de Brange's lemma for a convex cone of a real locally convex lattice of type (M).

Theorem 4.5.5 *Let E be a real locally convex lattice of type (M), $\mathcal{C} \subset E$ a convex cone, $A \subset ReZ(E)$ a subset containing 0 and W a convex, solid neighbourhood of the origin which is also a sublattice. If $f \in Ext\left\{W^0 \cap \mathcal{C}^0\right\}$ and $I = \{x \in E; |f|(|x|) = 0\}$, then $I \in \widetilde{A}_{A,\mathcal{C}}(E)$.*

Proof Let $U \in A$ with the properties:

$$(i) \, 0 \leq U_I \leq 1_{E/I},$$

$$(ii) \, U_I[\pi_I(\mathcal{C})] + [1_{E/I} - U_I][\pi_I(\mathcal{C})] \subset \pi_I(\mathcal{C}).$$

Since $f \in I^0 \subset E'$, it follows that there exists $g \in (E/I)'$ such that $f = \pi_I' g$.

Indeed, let $g : E/I \to \mathbb{R}$ given by

$$g\left[\pi_I(x)\right] = f(x), \, \forall x \in E.$$

We observe that g is well-defined since, if $\pi_I(x) = \pi_I(y)$, then $x - y \in I$, and taking into account that $f \in I^0$ we deduce that $f(x) = f(y)$.

Therefore, there exists $g \in (E/I)'$ such that

$$f = g \circ \pi_I = \pi_I' \circ g.$$

Obviously, we have

$$g \in Ext\left\{[\pi_I(W)]^0 \cap [\pi_I(\mathcal{C})]^0\right\}.$$

Furthermore, we introduce the following notations:

$$g_1 = U_I'g, \, g_2 = (1_{E/I} - U_I)'g$$

$$a_i = \inf\left\{\lambda > 0; \, g_i \in \lambda \cdot [\pi_I(W)]^0\right\} = \sup\{|g_i[\pi_I(x)]|; \, x \in W\}, i = 1, 2.$$

Since $g = g_1 + g_2 \in (a_1 + a_2) \cdot [\pi_I(W)]^0$, it follows that $a_1 + a_2 \geq 1$.

For any $x_1, x_2 \in W$, we have

$$|g_1[\pi_I(x_1)]| + |g_2[\pi_I(x_2)]| = |g|[U_I(\pi_I(x_1))]| + |g|[1_{E/I} - U_I](\pi_I(x_2))$$

$$\leq |g|[U_I(|\pi_I(x_1)| \vee |\pi_I(x_2)|) + (1_{E/I} - U_I)(|\pi_I(x_1)| \vee |\pi_I(x_2)|)]$$

$$= |g|(|\pi_I(x_1)| \vee |\pi_I(x_2)|).$$

We remark that $|\pi_I(x_1)| \vee |\pi_I(x_2)| \in \pi_I(W)$, since $\pi_I(W)$ is a solid set and also a sublattice.

Taking into account that $g \in [\pi_I(W)]^0$, we deduce that $|g|\,(|\pi_I(x_1)| \vee |\pi_I(x_2)|) \leq 1$, and so

$$|g_1[\pi_I(x_1)]| + |g_2[\pi_I(x_2)]| \leq 1, \ \forall x_1, x_2 \in W.$$

Therefore, we proved that $a_1 + a_2 \leq 1$, and so $a_1 + a_2 = 1$.

Furthermore, we shall prove that $g_i \neq 0$, $i = 1, 2$. First, we prove that

$$\left| U_I{'}(g) \right| = U_I{'}\,(|g|).$$

Indeed, since $0 \leq U_I{'} \leq 1_{(E/I)}{'}$ it follows that $U_I{'}(g_+) = (U_I{'}g)_+, U_I{'}(g_-) = (U_I{'}g)_-$, and so

$$\left| U_I{'}(g) \right| = U_I{'}(g)_+ + U_I{'}(g)_- = U_I{'}(g_+) + U_I{'}(g_-) = U_I{'}(g_+ + g_-) = U_I{'}(|g|).$$

If we suppose that $g_1 = 0$ and taking into account that $U \geq 0$, because $U_I \geq 0$, then for any $x \in E$ we have

$$|f|(|U(x)|) \leq |f|(U(|x|)) = |\pi_I{'}g|(U(|x|))$$

$$= |g|(U_I(\pi_I(|x|))) = |U_I{'}g|(\pi_I(|x|)) = |g_1|(\pi_I(|x|) = 0,$$

where from it results that $U(x) \in I$ and so that $U_I = 0$. Therefore, $I \in \mathcal{A}_{A,\mathcal{C}}(E)$.

Similarly, if we suppose that $g_2 = 0$ it results that $1_{E/I} - U_I = 0$ and so that $I \in \mathcal{A}_{A,\mathcal{C}}(E)$.

Therefore, we can suppose that $g_i \neq 0$, $i = 1, 2$, where from it we deduce that

$$a_i > 0, \ i = 1, 2.$$

Furthermore, we have

$$g = a_1 \cdot \frac{g_1}{a_1} + a_2 \cdot \frac{g_2}{a_2} \ \text{and} \ \frac{g_i}{a_i} \in [\pi_I(W)]^0, i = 1, 2.$$

On the other hand, since $g \in [\pi_I(\mathcal{C})]^0$, for any $x \in \mathcal{C}$ we have

$$\frac{g_1}{a_1} \in [\pi_I(W)]^0 \cap [\pi_I(\mathcal{C})]^0.$$

Similarly, we can prove that

$$\frac{g_2}{a_2} \in [\pi_I(W)]^0 \cap [\pi_I(C)]^0.$$

Taking into account that $g = a_1 \cdot \frac{g_1}{a_1} + a_2 \cdot \frac{g_2}{a_2}$ and $g \in Ext\left\{[\pi_I(W)]^0 \cap [\pi_I(C)]^0\right\}$, it follows that

$$g = \frac{g_1}{a_1} \text{ or } g = \frac{g_2}{a_2}.$$

We observe now that the equality $g_1 = a_1 \cdot g$ implies $(a_1 \cdot 1_{E/I} - U_I)' = 0$ and so

$$U_I = a_1 \cdot 1_{E/I}.$$

Similarly, the equality $g_2 = a_2 \cdot g$ implies $U_I = a_2 \cdot 1_{E/I}$ and therefore, in both cases, $I \in \mathcal{A}_{A,C}(E)$ and so the proof is finished.   □

Corollary 4.5.6 *Let* $E = CV_0(X, \mathbb{R})$, $A \subset C_b X, \mathbb{R})$ *be a subset containing* 0, $C \subset CV_0(X, \mathbb{R})$ *a convex cone, and* $L = v \cdot \mu \in Ext\left\{B_v{}^0 \cap C^0\right\}$.
   Then, the support of L *is an antialgebraic subset with respect to the pair* (A, C).

Proof If we denote by

$$I = \{f \in CV_0(X, \mathbb{R}); |L|(|f|) = 0\},$$

then, from the previous theorem, it follows that I is an antisymmetric ideal with respect to the pair (A, C). On the other hand, if we denote by S_L the support of L, then we have

$$I = \{f \in CV_0(X, \mathbb{R}); f|_{S_L} = 0\}.$$

Now, from Remark 4.5.3, we deduce that S_L is an antialgebraic subset with respect to (A, C).   □

Theorem 4.5.7 *Let* E *be a real locally convex lattice of* (M)*-type,* $A \subset Re[Z(E)]$ *a subset containing* 0, *and* $C \subset E$ *a convex cone. Then, for any* $x \in E$, *we have*

$$x \in \overline{C} \Leftrightarrow \pi_I(x) \in \overline{\pi_I(C)}, \forall I \in \widetilde{\mathcal{A}}_{A,C}(E),$$

where $\widetilde{\mathcal{A}}_{A,C}(E)$ *denotes the family of all minimal antisymmetric closed ideals with respect to the pair* (A, C).

Proof The necessity of the assertion is obvious. To prove the sufficient part, we assume that $\pi_I(x) \in \overline{\pi_I(C)}$, $\forall I \in \tilde{\mathcal{A}}_{A,C}(E)$ and $x \notin \overline{C}$. Then, there exists $f \in E'$ and $\varepsilon > 0$ such that

$$f(y) \leq 0 < \varepsilon < f(x), \ \forall y \in C.$$

On the other hand, from the continuity of f, it results that there exists a convex and solid neighbourhood of origin W which is also a sublattice such that

$$|f| \ (|x|) \leq 1, \ \forall x \in W.$$

Since $f \in C^0$, we deduce that $f \in C^0 \cap W^0$. From Krein-Milman theorem, we can suppose that $f \in Ext \left\{ W^0 \cap C^0 \right\}$. If we denote now by

$$J = \left\{ y \in E; |f| (|y|) = 0 \right\},$$

then, according to Theorem 4.5.5, it follows that $J \in \mathcal{A}_{A,C}E)$. On the other hand, from Corollary 4.5.4, it results that there exists an ideal $J_0 \in \tilde{\mathcal{A}}_{A,C}(E)$ such that $J_0 \subset J$.

Since by hypothesis $\pi_{J_0}(x) \in \overline{\pi_{J_0}(C)}$ and $f \in J_0{}^0 \cap C^0$, it follows that $f(x) \leq 0$. Indeed, let $g : E/J_0 \to \mathbb{R}$ given by $g[\pi_{J_0}(x)] = f(x)$. We remark that f is well-defined, since $f \in J_0{}^0$. Also, we observe that $f \in C^0$ implies that $g \in [\pi_{J_0}(C)]^0$ and furthermore, that $f(x) = g[\pi_{J_0}(x)] \leq 1$, and this is a contradiction because $f(x) > 0$. □

Corollary 4.5.8 *Let $E = CV_0(X, \mathbb{R})$, $\mathcal{F} \subset C(X, [0, 1])$ be a subset containing the constant functions and $C \subset CV_0(X, \mathbb{R})$ be a convex cone containing the constant function 0. Then*

$$\overline{C} = \left\{ g \in CV_0(X, \mathbb{R}); g \left| B_x \in \overline{C \left| B_x}, \forall x \in X \right. \right\},$$

where $\{B_x\}_{x \in X}$ denotes the family of all maximal $(\mathcal{F}, C)$-antialgebraic subsets of X.

The assertion follows from Theorem 4.5.7 and Remark 4.5.3.

Theorem 4.5.9 *Let E be a real locally convex lattice of (M)-type, $A \subset \{U \in Re(Z);\ 0 \leq U \leq 1_{E/I}\}$ and let $C \subset E_+$ be a convex cone with the properties:*

(i) *$A\,C = \{U(C);\ U \in A\} \subset C$,*

(ii) *C is not included in any maximal ideal of E,*

(iii) *Any closed ideal $I \subset E$ such that $\{U_I;\ U \in A\} \subset \mathbb{R}_+ \cdot 1_{E/I}$ is a maximal ideal. Then,*

$$\overline{C} = E_+.$$

Proof Let $x \in E_+$ and $I \in \tilde{A}_{A,C}(E)$. From hypothesis (i), it follows that for any $U \in A$ we have

$$U_I[\pi_I(\mathcal{C})] + [1_{E/I} - U_I][\pi_I(\mathcal{C})] \subset \pi_I(\mathcal{C}).$$

Since I is an $(A, \mathcal{C})$-antisymmetric ideal, there exists a real constant $a_U > 0$ such that

$$U_I = a_U \cdot 1_{E/I}.$$

Taking (iii) into account, we deduce that I is a maximal ideal. Hence the quotient space E/I has dimension 1. On the other hand, from hypothesis (ii), it follows that $\pi_I(\mathcal{C}) \neq \{0\}$. Now, from the inclusions $\{0\} \subset \pi_I(\mathcal{C} - \mathcal{C}) \subset \pi_I(E_+ - E_+) = \pi_I(E)$, we deduce that $\pi_I(\mathcal{C}) = \pi_I(E_+)$ and so that $\pi_I(x) \in \pi_I(\mathcal{C})$. Since $I \in \tilde{A}_{A,F}(E)$ was arbitrary, we deduce using Theorem 4.5.7 that $x \in \bar{\mathcal{C}}$, and so the proof is finished. □

In the particular case of the weighted spaces, we obtain the following corollary.

Corollary 4.5.10 *Let X be a Hausdorff locally compact space, V be a Nachbin family on X, $\mathcal{F} = C(X; [0, 1])$ and $\mathcal{C} \subset CV_0(X, \mathbb{R}_+)$ a convex cone such that $\mathcal{F} \cdot \mathcal{C} \subset \mathcal{C}$.*

We suppose, in addition, that $\mathcal{F}$ separates the points of X and for any $x \in X$ there exists a function $h_x \in \mathcal{C}$ such that $h_x(x) > 0$. Then,

$$\bar{\mathcal{C}} = CV_0(X, \mathbb{R}_+).$$

The assertion follows from Remark 4.5.3 and the prevision theorem.

4.6   Complex Locally Convex Lattices: Preliminaries and Notations

First, we shall present the main results and notions concerning the complex locally convex lattices. Our exposition is based on the following papers: D. Vuza [44], W.A. Luxemburg and A.C. Zaanen [17] and W.J. Schipper [39].

Definition 4.6.1 A complex vector lattice (c.v.l.) is a pair (E, m), where E is a complex vector space and m is a modulus mapping, i.e., $m : E \to E$ and has the following properties:

(i) $m(m(x)) = m(x), \forall x \in E$,
(ii) $m(\alpha \cdot x) = |\alpha| \cdot m(x), \forall \alpha \in \mathbb{C}, \forall x \in E$,
(iii) $m(m(m(x) + m(y)) - m(x + y)) = m(x) + m(y) - m(x + y), \forall x, y \in E$,
(iv) $E = Sp(m(E))$,

where $Sp(m(E))$ denotes the complex vector space generated by $m(E)$.

Proposition 4.6.2 *If E is a c.v.l., then $m(E)$ is a convex cone and $E_{\mathbb{R}} = m(E) - m(E)$ is a real vector lattice with respect to the order relation:*

$$x \leq y \Leftrightarrow y - x \in m(E).$$

Moreover, $x \geq 0 \Leftrightarrow x = m(x)$ and for any $x \in E_{\mathbb{R}}$ we have

$$m(x) = |x| = x \vee (-x).$$

Proposition 4.6.3 *If E is a c.v.l., then any $x \in E$ can be written in a unique way as the sum:*

$$x = x_1 + i \cdot x_2, \ x_i \in E_{\mathbb{R}}, \ i = 1.2.$$

If we denote by $Rex = x_1$ and $Imx = x_2$, then

$$x = Rex + i \cdot Imx.$$

Also, we shall denote

$$x^* = Rex - i \cdot Imx.$$

Definition 4.6.4 Let E be a c.v.l. A complex vector sublattice of E is a complex vector subspace $F \subset E$ with the following property:

$$x \in F \text{ implies } Rex \in F \text{ and } m(x) \in F.$$

A subset $S \subset E$ is said to be solid, if it has the property:

$$\forall x \in S, \ \forall y \in E \text{ such that } m(y) \leq m(x) \text{ involves } y \in S.$$

An ideal $I \subset E$ is a complex vector subspace of E which is also a solid subset.

Proposition 4.6.5 *Let E be a c.v.l., $I \subset E$ be a closed ideal and $\pi_I : E \to E/I$ the canonical mapping. The modulus on the quotient vector space E/I is defined by*

$$\widehat{m_I}\left[\pi_I(x)\right] = \pi_I\left[m(x)\right], \ \forall x \in E.$$

The pair $(E/I, \ \widehat{m_I})$ is also a c.v.l.

Definition 4.6.6 Let E be a c.v.l. A seminorm $q : E \to \mathbb{R}$ is called a solid seminorm if

$$m(x) \le m(y) \implies q(x) \le q(y), \ \forall x, y \in E.$$

Moreover, for any solid seminorm q, we have

$$q(x) = q\,[m(x)], \ \forall x \in E.$$

Proposition 4.6.7 *Let E be a c.v.l., $q : E \to \mathbb{R}_+$ a solid seminorm and $J = \{x \in E; q(x) = 0\}$. Then, J is an ideal and $\widehat{q} : E/J \to \mathbb{R}_+$ defined by $\widehat{q}\,[\pi_J(x)] = q(x), \ \forall x \in E$ is a solid norm on E/J.*

Definition 4.6.8 A c.v.l. E is called Archimedean if $E_{\mathbb{R}}$ is Archimedean (i.e., $x \le 0$ whenever $\alpha \cdot x \le y$ for some $y \in E_+$ and all $\alpha > 0$).

Definition 4.6.9 A c.v.l. with strong unit is an Archimedean c.v.l. E, endowed with a solid norm $\| \| : E \to \mathbb{R}$ such that there exists an element $e \in E_+$ (the strong unit of E) with the properties:

$$\|e\| = 1 \text{ and } m(x) \le \|x\| \cdot e, \forall x \in E$$

Remark 4.6.10 Let E be a c.v.l. For any $x \in E$ we shall denote by E_x the principal ideal generated by x, i.e.,

$$E_x = \{y \in E; \ \exists \lambda \in \mathbb{R}_+, \ \text{such that } m(y) \le \lambda \cdot m(x)\}.$$

Obviously, E_x is a (complex) ideal of E and $q_x : E_x \to \mathbb{R}_+$ given by

$$q_x(y) = \inf\{\lambda > 0 \ ; \ m(y) \le \lambda \cdot m(x)\}, \forall y \in E_x,$$

is a solid seminorm on E_x. If we denote by J_x the kernel of q_x, then E_x/J_x is a c.v.l. with strong unit, with respect to the norm $\widehat{q}_x$ and the unit $\pi_{J_x}[m(x)]$.

Theorem 4.6.11 *Let E be a c.v.l. with strong unit e. Then, there exists a Hausdorff compact space X and an injective lattice morphism $T : E \to C(X, \mathbb{C})$ (i.e., $|T(x)| = T(m(x)), \forall x \in E$) such that*

$$T(e) = 1 \text{ and } \overline{T(E)} = C(X, \mathbb{C})$$

Proposition 4.6.12 *Let E be a c.v.l. Then, for any $x \in E$ we have*

$$m(x) = \sup \{ Re(a \cdot x); a \in C, |a| = 1 \},$$

$$Re(x) \leq m(x), \, Im(x) \leq m(x).$$

Definition 4.6.13 Let E be a c.v.l., and $r \in E_+$. A sequence $\{x_n\}_n$, $x_n \in E$, $\forall n \in \mathbb{N}^*$, is called r-uniformly convergent to $x \in E$, if for any $\forall \varepsilon > 0$, there exists a natural number n_ε such that

$$m(x_n - x) < \varepsilon \cdot r, \forall n \geq n_\varepsilon.$$

The sequence $\{x_n\}$ is called r-uniformly fundamental if $\forall \varepsilon > 0$, there exists $n_\varepsilon \in \mathbb{N}^*$ such that

$$m(x_{n+p} - x_n) < \varepsilon \cdot r, \forall n \geq n_\varepsilon, \forall p \in \mathbb{N}^*.$$

A c.v.l. is called relatively uniformly complete if for any $r \in E_+$ and any r-uniformly fundamental sequence $\{x_n\}$, there exists $x \in E$ such that the sequence $\{x_n\}$ is r-uniformly convergent to x.

Proposition 4.6.14 *Every Archimedean and relatively uniformly complete c.v.l. has the Riesz decomposition property, i.e. $\forall x \in E$, and $\forall y_1, y_2 \in E_+$ with the property $m(x) \leq y_1 + y_2$, there exist $x_1, x_2 \in E$ such that*

$$x = x_1 + x_2 \text{ and } m(x_1) \leq y_1, \, m(x_2) \leq y_2$$

Definition 4.6.15 Let E_1 and E_2 be two c.v.l. A (complex) linear operator $T : E_1 \rightarrow E_2$ is called order bounded if T maps order bounded sets into order bounded sets (i.e., for any $x \in (E_1)_+$ there exists $u \in (E_2)_+$ such that $m_1(y) \leq x$ implies $m_2[T(y)] \leq u$).

Proposition 4.6.16 *Let E be an Archimedean and relatively uniformly complete c.v.l., and let F be an order complete c.v.l. Then, any order bounded complex linear operator $T : E \rightarrow F$ has a linear modulus $|T|$ given for any $x \in E_+$ by*

$$|T|(x) = \sup \{ m(T(y)); m(y) \leq x \} = \sup \{ Re(\lambda \cdot T)(x); \lambda \in C, |\lambda| = 1 \}$$

($|T|$ is additive on E_+ and can be extended to a positive linear operator from E to F).

Definition 4.6.17 A complex locally convex lattice of type (M) is a Hausdorff relatively uniformly complete c.v.l. endowed with a linear topology such that there exists a neighbourhood basis of the origin consisting of sublattices, which are also solid and convex.

Definition 4.6.18 Let E be a c.v.l. A seminorm $p : E \to \mathbb{R}$ is called of type (M) if it has the property:

$$p(x \vee y) = \max{(p(x), p(y))}, \ \forall x, y \in E_+.$$

Proposition 4.6.19 *If (E, τ) is a relatively uniformly complete c.v.l. and τ is a Hausdorff locally convex topology on E, then E is a complex locally convex lattice of type (M) iff the topology τ is defined by a set of solid seminorms of type (M) (see [9, p. 198]).*

The typical example of complex locally convex lattice of type (M) is the weighted space $CV_0(X, \mathbb{C})$.

Remark 4.6.20 Every complex locally convex lattice E has the property:

- For any $x \in E$, $x \neq 0$ there exists $f \in E'$, $f(x) \neq 0$.

Here, E' denoted the dual of E, i.e., the space of all complex valued linear and continuous functions on E. Since any $f \in E'$ is order bounded, we infer that for every complex locally convex lattice E there exists a sufficient set of order bounded functionals on E.

4.7   Antisymmetric Ideals in Complex Locally Convex Lattices of (M)-Type

Let E be a complex locally convex lattice of type (M). The center $Z(E)$ of E is the algebra of all order bounded endomorphisms on E, that is, those operators $U \in L(E, E)$ for which there exists $\lambda_U > 0$ such that

$$m[U(x)] \leq \lambda_U \cdot m(x), \ \forall x \in E.$$

The real part of the center is denoted by $ReZ(E)$ and is defined by

$$ReZ(E) = Z(E)_+ - Z(E)_+.$$

For any closed ideal I of E and any $U \in Z(E)$, we define $U_I : E/I \to E/I$ by

$$U_I[\pi_I(x)] = \pi_I[U(x)], \ \forall x \in E.$$

It is easy to prove that U_I is well-defined and $U_I \in Z(E/I)$.

Definition 4.7.1 Let I and J be two closed ideals of E such that $I \subset J$. Then, we can define the following two mappings:

$$\pi_{IJ} : E/I \to E/J, \ \pi_{IJ}[\pi_I(x)] = \pi_J(x), \ \forall x \in E,$$

$$M_{IJ} : Z(E/I) \to Z(E/J), \ M_{IJ}(T)[\pi_J(x)] = \pi_{IJ}(T[\pi_I(x)]), \ \forall T \in Z(E/I)$$

Proposition 4.7.2 *The mappings π_{IJ} and M_{IJ} have the following properties:*

$(i) \pi_{IJ}$ *is a lattice homomorphism,*

$(ii) \ M_{IJ}(T) \in Z(E/J), \ \forall T \in Z(E/I)$

$(iii) \ M_{IJ}(T) \in Z(E/J)_+, \ \forall T \in Z(E/I)_+$

Proof

(i) For any $x \in E$ we have

$$\widehat{m_J}(\pi_{IJ}[\pi_I(x)]) = \widehat{m_J}[\pi_J(x)] = \pi_J[m(x)] = \pi_{IJ}(\pi_I[m(x)]) = \pi_{IJ}(\widehat{m_I}[\pi_I(x)]).$$

Therefore,

$$\widehat{m_J}(\pi_{IJ}[\pi_I(x)]) = \pi_{IJ}(\widehat{m_I}[\pi_I(x)]).$$

(ii) For any $T \in Z(E/I)$, we have

$$\widehat{m_J}([M_{IJ}(T)][\pi_J(x)]) = \widehat{m_J}(\pi_{IJ}(T[\pi_I(x)])) = \pi_{IJ}(\widehat{m_I}[T(\pi_I(x))])$$
$$\leq \pi_{IJ}(\lambda_T \cdot \widehat{m_I}[\pi_I(x)]) = \lambda_T \cdot \pi_{IJ}[\pi_I(\widehat{m_I}(x))]$$
$$= \lambda_T \cdot \pi_{IJ}[\pi_I(m(x)] = \lambda_T \cdot \pi_J[m(x)] = \lambda_T \cdot \widehat{m_J}[\pi_J(x)]$$

(iii) If $\pi_J(x) \geq 0$, we have

$$\pi_J(x) = \widehat{m_J}[\pi_J(x)].$$

Furthermore, for any $T \in Z(E/I)_+$ it results

$$\widehat{m_J}(M_{IJ}(T)[\pi_J(x)]) = \widehat{m_J}(M_{IJ}(T)[\pi_J(m(x))])$$
$$= \widehat{m_J}(\pi_{IJ}[T(\pi_I(m(x)))]) = \pi_{IJ}(\widehat{m_I}[T(\pi_I(m(x)))])$$
$$= \pi_{IJ}[T(\pi_I(m(x))] = M_{IJ}(T)[\pi_J(m(x))] = M_{IJ}(T)[\pi_J(x)],$$

Hence $M_{IJ}(T) \geq 0$.

$\square$

Definition 4.7.3 Let $A \subset Z(E)$ be a subset containing the constant operator 0 and $F \subset E$ a complex vector subspace. A closed ideal $I \subset E$ is said to be antisymmetric with respect to the pair (A, F) (in brief (A, F)-antisymmetric) if for any $U \in A$ with the properties:

$$\text{(i)}\ \ U_I \in ReZ(E/I)$$
$$\text{(ii)}\ \ U_I[\pi_I(F)] \subset \pi_I(F),$$

there exists some $a \in R$, such that

$$U_I = a \cdot 1_{E/I},$$

where $1_{E/I}$ is the identity operator on the quotient space E/I.

Obviously, the lattice E itself is antisymmetric with respect to the pair (A, F).

Furthermore, we shall denote by $\mathcal{A}_{A,F}(E)$ the family of all closed ideals of E antisymmetric with respect to the pair (A, F).

Proposition 4.7.4 *Let X be a Hausdorff locally compact space, and let V be a Nachbin family on X. Then, the center of the corresponding weighted space $CV_0(X, \mathbb{C})$ coincides with the Banach algebra $C_b(X, \mathbb{C})$ of all bounded continuous functions on X.*

If K is a compact Hausdorff space, then the center of $C(X, \mathbb{C})$ is $C(X, \mathbb{C})$ itself.

The proof is similar to that of Proposition 4.3.4, replacing $\mathbb{R}$ by $\mathbb{C}$ when it is necessary.

Theorem 4.7.5 *Let $I \subset CV_0(X, \mathbb{C})$ be a closed ideal. Then, there exists a closed subset $S_I \subset X$ such that*

$$I = \{f \in CV_0(X, \mathbb{C});\ f|S_I = 0\}.$$

Proof Let $I \subset CV_0(X, \mathbb{C})$ be a closed ideal and let $J = I \cap CV_0(X, \mathbb{R})$. Clearly, J is a closed ideal of $CV_0(X, \mathbb{R})$. By Theorem 4.2.1, it results that there exists a closed subset $S_I \subset X$ such that

$$J = \{f \in CV_0(X, \mathbb{R});\ f|S_I = 0\}.$$

On the other hand, we have $I = J + i \cdot J$, and therefore

$$I = \{f \in CV_0(X, \mathbb{C});\ f|S_I = 0\}. \qquad \square$$

Definition 4.7.6 Let $A \subset C_b(X, \mathbb{C})$ be a subset containing the constant function 0 and $F \subset CV_0(X, \mathbb{C})$ be a complex vector subspace. A subspace $S \subset X$ is called antisymmetric with respect to the pair (A, F) (in brief (A, F)-antisymmetric) if any function $a \in A$ such that

$$\text{(i) } a(S) \subset \mathbb{R},$$

$$\text{(ii) } a \cdot f \,|S \in F \,|S \,, \forall f \in F,$$

is constant on S.

Remark 4.7.7 A closed subset $S \subset X$ is (A, F)-antisymmetric, in the sense of Definition 4.7.6. if and only if the closed ideal $I_S = \{f \in CV_0(X, \mathbb{C}; \, f \,|S = 0\}$ is (A, F)-antisymmetric in the sense of Definition 4.7.3.

Indeed, according to Proposition 4.7.4, the center of $CV_0(X, \mathbb{C})$ is isomorphic with $C_b(X, \mathbb{C})$. Moreover, from the proof, it results that the isomorphism $U \in Z[CV_0(X, \mathbb{C})] \leftrightarrow a_U \in C_b(X, \mathbb{C})$ consists in the equality:

$$U(f) = a_U \cdot f, \, \forall f \in CV_0(X, \mathbb{C}).$$

On the other hand, for any $h \in CV_0(X, \mathbb{C})$ we have $\pi_{I_S}(h) = h \,|S$ and further

$$U_{I_S}[\pi_{I_S}(f)] = \pi_{I_S}(U(f)) = \pi_{I_S}(a_U \cdot f) = (a_U \cdot f) \,|S \,.$$

Therefore, $U \in A$ and it has the property:

$$U_{I_S}[\pi_{I_S}(F)] \subset \pi_{I_S}(F) \quad \Leftrightarrow \quad a_U \cdot F \,|S \subset F \,|S \,.$$

We also remark that $U_{I_S} = a \cdot 1_{CV_0(X, \mathbb{C})/I_S}$ is equivalent to the fact that $a_U(x) = a, \, \forall x \in S$, and so a_U is constant on S.

Remark 4.7.8 Let K be a compact Hausdorff space, and let $\mathcal{A} \subset C(X, \mathbb{C})$ be a subalgebra. A closed subset $S \subset K$ is called antisymmetric with respect to algebra $\mathcal{A}$ if any $a \in \mathcal{A}$ such that $f(x) \in \mathbb{R}, \, \forall x \in S$ is constant on S.

This is the original definition of an antisymmetric set with respect to a subalgebra $\mathcal{A} \subset C(X, \mathbb{C})$ introduced by E.Bishop [2] (see also Glicksberg [10]). Obviously, this definition follows from Definition 4.7.6 for $A = F = \mathcal{A}$.

Now, we come back to the general case of antisymmetric ideals of a complex locally convex lattice of type (M).

Theorem 4.7.9 *Let E be a complex locally convex lattice of type (M), $A \subset Z(E)$ be a subset containing the constant operator 0 and $F \subset E$ be a complex vector subspace.*

If $I_\alpha \in \mathcal{A}_{A,F}(E)$, $\forall \alpha$ and $J = \sum_\alpha I_\alpha \neq E$, then

$$I = \bigcap_\alpha I_\alpha \in \mathcal{A}_{A,F}(E).$$

The proof is similar to the proof of Theorem 4.3.7, taking also into account Proposition 4.7.2.

Corollary 4.7.10 *For any closed ideal $I \in \mathcal{A}_{A,F}(E)$, there exists a unique minimal closed ideal $\tilde{I} \in \mathcal{A}_{A,F}(E)$, such that $\tilde{I} \subset I$.*

Proof Let $I \in \mathcal{A}_{A,F}(E)$, $I \neq E$ and let $\tilde{I} = \cap \{J \in \mathcal{A}_{A,F}(E); J \subset I\}$. From Theorem 4.7.9, it results that $\tilde{I} \in \mathcal{A}_{A,F}(E)$. On the other hand, obviously $\tilde{I}$ is minimal.

Furthermore, we shall denote by $\tilde{\mathcal{A}}_{A,F}(E)$ the family of all ideals antisymmetric with respect to the pair (A, F). □

4.8 The Generalization of de Brange Lemma for a Locally Convex Lattice of Type-(M)

First, we present two remarks concerning the positive linear operators (endomorphisms) on a locally convex lattice of type (M)

Remark 4.8.1 Let E be a complex lattice, and $U : E \to E$ a positive linear operator. Then,

$$m[U(x)] \leq U[m(x)], \ \forall x \in E.$$

Indeed, if $x \in E_{\mathbb{R}}$, then $U(x) \in E_{\mathbb{R}}$ and

$$m(U(x)) = U(x) \vee -U(x) \leq U(m(x)), \forall x \in E.$$

By Proposition 4.6.12, for any $x \in E$ we have

$$m(U(x)) = \sup_{\substack{a \in \mathbb{C} \\ |a|=1}} Rea \cdot U(x) = \sup_{\theta \in [0,2\pi]} (\cos\theta \cdot ReU(x) - \sin\theta \cdot ImU(x)) =$$

$$= \sup_{\theta \in [0,2\pi]} (\cos\theta \cdot U(Rex) - \sin\theta \cdot U(Imx)) = \sup_{\substack{a \in \mathbb{C} \\ |a|=1}} U(Re(a \cdot x)) \leq U(m(x)).$$

Remark 4.8.2 Let $U : E \to E$ be a positive linear operator and $g : E \to \mathbb{C}$ be an order bounded linear functional. Then,

$$|g\,[U(x)]| \leq |g|\,[U(m(x))]\,.$$

The assertion follows from Remark 4.8.1 and from the formula:

$$|g|\,[U(m(x))] = \sup\{|g\,(z)|\,;\ m(z) \leq U\,[m(x)]\}\,.$$

Theorem 4.8.3 *Let E be a c.v.l. of type (M), $A \subset Z(E)$ be a subset containing the constant operator 0, $F \subset E$ be a complex vector subspace, and W be a convex, solid neighbourhood of the origin which is also a sublattice. If $f \in Ext\{W^0 \cap F^0\}$ and*

$$I = \{x \in E;\ |f|\,(m(x)) = 0\}\,,$$

then I is antisymmetric with respect to the pair (A, F).

Proof *Let $U \in A$ be with the properties:*

(i) $U_I \in ReZ(E/I)$,

(ii) $U_I[\pi_I(F)] \subset \pi_I(F)$.

Since $U_I \in ReZ(E/I)$ and $ReZ(E/I) = ReZ(E/I)_+ - ReZ(E/I)_+$, we may suppose that

$$0 \leq U_I \leq 1_{E/I}.$$

Indeed, let $S, T \in Z(E/I)_+$ be such that $U_I = S - T$. Since S is an order bounded endomorphism on E/I, one can find $\lambda_S > 0$ such that

$$\widehat{m_I}\,[S(\pi_I(x))] \leq \lambda_S \cdot \widehat{m_I}\,[\pi_I(x)]\,, \ \forall x \in E,$$

and so

$$S(\pi_I(x)) \leq \lambda_S \cdot \pi_I(x), \forall \pi_I(x) \geq 0.$$

If we replace S by $\frac{1}{\lambda_S} \cdot S$, we get $0 \leq S \leq 1_{E/I}$. On the other hand, it is clear that $S[\pi_I(F)] \subset \pi_I(F)$.

Similarly, we can prove that $0 \leq T \leq 1_{E/I}$ and $T[\pi_I(F)] \subset \pi_I(F)$. If we prove that $S = a \cdot 1_{E/I}$ and $T = b \cdot 1_{E/I}$, then it will follow that $U_I = (a - b) \cdot 1_{E/I}$.

Therefore, further we may suppose

$$U_I \in Z(E/I)_+, \quad 0 \leq U_I \leq 1_{E/I}, \quad U_I[\pi_I(F)] \subset \pi_I(F).$$

Since $f \in I^0 \subset E'$, it follows that there exists $g \in (E/I)'$ such that $f = \pi_I(g)$.
Indeed, let $g : E/I \to \mathbb{R}$ be given by

$$g[\pi_I(x)] = f(x), \ \forall x \in E.$$

We observe that g is well-defined because, if $\pi_I(x) = \pi_I(y)$, then $x - y \in I$ and taking
into account that $f \in I^0$, it results that $f(x) = f(y)$. Therefore, there exists $g \in (E/I)'$
such that

$$f = g \circ \pi_I = \pi_I{}' \circ g.$$

Obviously, we have

$$g \in Ext\left\{ [\pi_I(W)]^0 \cap [\pi_I(F)]^0 \right\}.$$

Furthermore, we shall denote

$$g_1 = U_I{}' \circ g, \quad g_2 = (1_{E/I} - U)' \circ g,$$

and

$$a_i = \inf\left\{ \lambda > 0; \ g_i \in \lambda \cdot [\pi_I(W)]^0 \right\} = \sup\left\{ |g_i[\pi_I(x)]| \, ; \, x \in W \right\}, i = 1, 2.$$

Since $g = g_1 + g_2 \in (a_1 + a_2) \cdot [\pi_I(W)]^0$, we get $a_1 + a_2 \geq 1$.
On the other hand, for any $x_1, x_2 \in W$, we have

$$|g_1[\pi_I(x_1)]| + |g_2[\pi_I(x_2)]| = |g\left(U_I[\pi_I(x_1)]\right)| + \left|g(1_{E/I} - U_I)[\pi_I(x_2)]\right|.$$

If we denote by

$$y = \hat{m} \, [\pi_I(x_1)] \vee \hat{m} \, [\pi_I(x_2)],$$

and, taking into account that $\pi_I(W)$ is a solid neighbourhood of the origin which is also a
sublattice, it follows that $y \in \pi_I(W)$.

Now, from Remark 4.8.2, we deduce

$$|g_1[\pi_I(x_1)]| + |g_2[\pi_I(x_2)]| \le |g|[U_I(y)] + |g|[(1_{E/I} - U_I)(y)]$$

$$= |g|[(U_I)(y) + y - U_I(y)] = |g|(y).$$

Since $g \in [\pi_I(W)]^0$, it follows $|g|(y) \le 1$. Therefore, we have

$$|g_1[\pi_I(x_1)]| + |g_2[\pi_I(x_2)]| \le 1, \forall x_1, x_2 \in W,$$

hence $a_1 + a_2 \le 1$ and so $a_1 + a_2 = 1$. Furthermore, we shall prove that $g_i \ne 0$, $i = 1, 2$.
First, we prove that $\left| U_I'(g) \right| \le U_I'(|g|)$.

Indeed, since $0 \le U_I' \le 1_{(E/I)}$, we have $U_I'(g_+) = U_I'(g)_+$; $U_I'(g_-) = U_I'(g)_-$, and further

$$\left| U_I'(g) \right| = U_I'(g)_+ + U_I'(g)_- = U_I'(g_+) + U_I'(g_-) = U_I'(g_+ + g_-) = U_I'(|g|).$$

Now, from Remark 4.8.1, it follows

$$|f|(m(U(x))) \le |f|(U(m(x))) = |\pi_I'g|(U(m(x)) = \pi_I'|g|(U_I(\pi_I(m(x)))$$

$$= |g|(U_I(\pi_I(m(x)))) = |U_I'g|(\pi_I(m(x))).$$

If we suppose that $g_1 = 0$, and taking into account that $U \ge 0$ (since $U_I \ge 0$), then it follows that

$$|f|(m(U(x))) \le \left| U_I'g \right|(\pi_I(m(x))) = |g_1|(\pi_I(m(x))) = 0.$$

Therefore, $U(x) \in I$ and so $U_I = 0$. Thus $g_1 = 0$ implies $U_I = 0 = 0 \cdot 1_{E/I}$, from which we deduce that $I \in \mathcal{A}_{A,F}(E)$. Using a similar argument, we prove that $g_2 = 0$ implies that $1_{E/I} - U_I = 0$, hence $U = 1 \cdot 1_{E/I}$ and so $\mathcal{A}_{A,F}(E)$. Therefore, we can further suppose that $g_i \ne 0$, $i = 1, 2$.
On the other hand, since $g \in [\pi_I(F)]^0$ and $U_I[\pi_I(F)] \subset \pi_I(F)$, we deduce that $g_i \in [\pi_I(F)]^0$, $i = 1, 2$. and so that $\frac{g_i}{a_i} \in [\pi_I(F)]^0 \cap [\pi_I(W)]^0$, $i = 1, 2$. Taking into account that $g = a_1 \cdot \frac{g_1}{a_1} + a_2 \cdot \frac{g_2}{a_2}$, and g is an extremal element of the set $[\pi_I(W)]^0 \cap [\pi_I(F)]^0$, we deduce that either $g = \frac{g_1}{a_1}$ or $g = \frac{g_2}{a_2}$. We observe now that from the equality $g = \frac{g_1}{a_1}$ it results $\left(a_1 \cdot 1_{E/I} - U_I \right)' = 0$, hence $U_I = a_1 \cdot 1_{E/I}$. By a similar argument, the equality $g = \frac{g_2}{a_2}$ involves $U_I = a_2 \cdot 1_{E/I}$ and this concludes the proof.   □

Corollary 4.8.4 *Let $E = CV_0(X, \mathbb{C})$, $A \subset C_b(X, \mathbb{C})$ be a subset containing the constant function 0 and let F be a complex vector subspace of $CV_0(X, \mathbb{C})$. If $L = v \cdot \mu \in$*

$Ext\left\{B_v^0 \cap F^0\right\}$, *then the support of L is an antisymmetric subset with respect to the pair* (A, F) *(in the sense of Definition 4.7.6).*

Proof If we denote

$$I = \{f \in CV_0(X, \mathbb{C}); |L|\,(|f|) = 0\},$$

then from the above theorem we deduce that I is an antisymmetric ideal with respect to the pair (A, F). On the other hand, if we denote by $S_L = \text{supp}(L)$, then we have

$$I = \{f \in CV_0(X, \mathbb{C}); \ f\,|S_L = 0\}.$$

From Remark 4.7.7 it follows now that S_L is an antisymmetric subset with respect to the pair (A, F). □

4.9   Approximation of the Elements of a Locally Convex Lattice of (M)-Type by Using One of Subspaces

The main result of this section is the following approximation theorem.

Theorem 4.9.1 *Let E be a complex locally convex lattice of type (M), $A \subset Z(E)$ be a subset containing the constant operator 0 and F be a complex vector subspace of E. Then, for any $x \in E$, we have*

$$x \in \overline{F} \Leftrightarrow \pi_I(x) \in \overline{\pi_I(F)}, \forall I \in \tilde{\mathcal{A}}_{A,F}(E),$$

where we denote by $\tilde{\mathcal{A}}_{A,F}(E)$ the family of all minimal (A, F)-antisymmetric ideals.

Proof The necessity is obvious. Conversely, if we suppose that $\pi_I(x) \in \overline{\pi_I(F)}$, $\forall I \in \tilde{\mathcal{A}}_{A,F}(E)$ and $x \notin \overline{F}$, then there exists a continuous linear functional $f \in E'$ such that $f(x) \neq 0$ and $f(y) = 0$, $\forall y \in F$. From the continuity of f, it results that there exists a convex, solid neighbourhood of the origin W which is also a sublattice such that

$$|f|\,[\,m(x)\,] \leq 1, \forall x \in W.$$

Since $f \in F^0$, we have $f \in W^0 \cap F^0$. From Krein-Milman theorem, we may assume that $f \in Ext\left\{W^0 \cap F^0\right\}$. If we denote

$$J = \{y \in E; |f|\,(m(x)) = 0\},$$

then, according to Theorem 4.8.3, it follows that J is a (A, F)-antisymmetric ideal. On the other hand, by Corollary 4.7.10, there exists a minimal (A, F)-antisymmetric ideal $J_0 \subset J$.

Since $f \in J_0{}^0 \cap F^0$ and, by hypothesis, $\pi_{J_0}(x) \in \overline{\pi_{J_0}(F)}$, we deduce that $f(x) = 0$. Indeed, let $g : E/J_0 \to \mathbb{R}$, given by $g[\pi_{J_0}(x)] = f(x)$. We observe that g is well-defined because $f \in J_0^0$.

Since $f \in F^0$, it follows that $g \in [\pi_{J_0}(F)]^0$ and taking into account that $\pi_{J_0}(x) \in \overline{\pi_{J_0}(F)}$, we deduce that $f(x) = g[\pi_{J_0}(x)] = 0$, and this contradicts the choice of f. $\square$

Remark 4.9.2 Let E, A and F be as in Definition 4.7.3, with the additional hypothesis:

$$AF = \{U(x); \ \forall U \in A, \ \forall x \in F\} \subset F.$$

Since for any $U \in A$ and any closed ideal $I \subset E$, we have $U_I[\pi_I(F)] \subset \pi_I(F)$, it follows that I is antisymmetric with respect to the pair (A, F) (or A-antisymmetric) if for any $U \in A$ with the property $U_I \in ReZ(E/I)$ there exists $a \in R$ such that $U_I = a \cdot 1_{E/I}$.

Furthermore, we shall denote by $\tilde{A}_A(E)$ the family of all A-antisymmetric closed ideals of E.

From Theorem 4.9.1 and Remark 4.9.2, the subsequent theorem follows.

Theorem 4.9.3 *Let E, A and F be as in Remark 4.9.2. Then, for any $x \in E$ we have*

$$x \in \overline{F} \text{ if and only if } \pi_I(x) \in \overline{\pi_I(F)} , \ \forall I \in \tilde{A}_A(E).$$

Remark 4.9.4 Let $A \subset C_b(X, \mathbb{C})$ be a subset containing the constant function 0 and $F \subset CV_0(X, \mathbb{C})$ a complex vector subspace such that $A \cdot F \subset F$. From Definition 4.7.6, Remarks 4.7.7 and 4.9.2, it results that a subset $S \subset X$ is antisymmetric with respect to the pair (A, F) (in brief A-antisymmetric) if any function $a \in A$ with the property $a(S) \subset R$ is constant on S.

Furthermore, for any $x \in X$, we denote

$$[x] = \{y \in X; \ a(y) = a(x), \ \forall a \in A\} .$$

Obviously, $[x]$ is the maximal A-antisymmetric subset of X containing the point x.

From Theorem 4.9.3 and Remark 4.9.4, we deduce the following Prolla's result [33].

Theorem 4.9.5 *Let E, A and F be as in Remark 4.9.4. Then, for any $f \in CV_0(X, \mathbb{C}) = E$, we have*

$$f \in \overline{F} \text{ iff } f \,|\, [x] \in \overline{F \,|\, [x]}, \ \forall x \in X.$$

Remark 4.9.6 If E is a complex vector lattice and $I \subset E$ is a maximal closed ideal, then the quotient space $E/I = \pi_I(E)$ has the dimension 2.

Indeed, if we denote by $J = I \cap E_\mathbb{R}$, then J is a maximal ideal in the real vector lattice $E_\mathbb{R}$ and it is known that J has the dimension 1. As $I = J + i \cdot J$, it follows that I has the dimension 2.

The next result is an abstract version of Stone–Weierstrass theorem for complex locally convex lattices of type (M).

Theorem 4.9.7 *Let E be a complex locally convex lattice of type (M), $A \subset Z(E)$ be a (complex) self-adjoint vector subspace, and $F \subset E$ be a complex vector space with the properties:*

(i) *$AF = \{U(x);\ \forall U \in A,\ \forall x \in F\} \subset F$.*
(ii) *F is not included in any maximal ideal of E.*
(iii) *Any closed ideal $I \subset E$ such that $\{U_I;\ U \in A\} \subset \mathbb{C} \cdot 1_{E/I}$ is a maximal ideal. Then,*

$$\overline{F} = E.$$

Proof Let $x \in E$ and $I \in \tilde{A}_{A,F}(E)$. Since $A \subset Z(E)$ is self-adjoint (i.e. $U = ReU + i \cdot ImU \in A$ implies that $U^* = ReU - i \cdot ImU \in A$), it results that $ReU,\ ImU \in A$ for any $U \in A$.

Taking into account the hypothesis (i) and the fact that I is (A, F)-antisymmetric, we deduce that there are $a, b \in R$ such that $(ReU)_I = a \cdot 1_{E/I}$ and $(ImU)_I = b \cdot 1_{E/I}$ for any $U \in A$. Therefore, we have $U_I = (a + i \cdot b) \cdot 1_{E/I}$ and so, by hypothesis (iii), I is a maximal ideal in E. From Remark 4.9.6, it follows that the quotient space E/I has the dimension 2.

As $\{0\} \subset \pi_I(F) \subset \pi_I(E) = E/I$, we have either $\pi_I(F) = \{0\}$ or $\pi_I(F) = \pi_I(E)$. On the other hand, by hypothesis (ii), it results that $\pi_I(F) \neq \{0\}$.

Therefore, $\pi_I(F) = \pi_I(E)$, hence $\pi_I(x) \in \pi_I(F),\ \forall I \in \tilde{A}_{A,F}(E)$. According to Theorem 4.9.1, it follows that $x \in \overline{F}$.   □

In the particular case of weighted spaces, we obtain the following Stone-Weierstrass theorem.

Theorem 4.9.8 *Let $A \subset C_b(X, \mathbb{C})$ be a self-adjoint algebra, and let $F \subset CV_0(X, \mathbb{C})$ be a complex vector subspace with the properties:*

(i) *$A \cdot F \subset F$*
(ii) *A separates the points of X*
(iii) *For any $x \in X$, there exists a function $f \in F$ such that $f(x) \neq 0$. Then, we have*

$$\overline{F} = CV_0(X, \mathbb{C}).$$

Proof From (iii), it results that F is not included in any maximal ideal of $CV_0(X, \mathbb{C})$. Indeed, for any maximal ideal $I \subset CV_0(X, \mathbb{C})$, there exists a point $x \in X$ such that

$$I = I_x = \{f \in CV_0(X, \mathbb{C}); \ f(x) = 0\}.$$

If we suppose that $F \subset I_x$, then $f(x) = 0$, $\forall f \in F$, and this contradicts hypothesis (iii).

From (i) and (ii), it follows that any (A, F)-antisymmetric subset $S \subset X$ is a singleton, i.e. $S = \{x\}$, where it follows that the corresponding ideal I_S is maximal.

The assertion follows now from Theorem 4.9.7. $\qquad\square$

Bibliography

1. Altomare, F., Campiti, M., *Korovkin –type Approximation Theory and Its Applications*, Walter de Gruyter, Berlin .New York 1994.
2. Bishop, E., A generalization of the Stone-Weierstrass theorem, Pac.J.Math. 11 (1961), 777–783.
3. Boboc, N., Bucur, Gh., Conuri convexe de functii continue pe spatii compacte, Ed.Academiei R.S.R., Bucuresti, 1976.
4. Branges, Louis de. The Stone-Weierstrass theorm, Proc.Amer.Math.Soc. 10, (1959), 822–824
5. Brosowski, B., Deutsch, F., An elementary proof of the Stone-Weierstrass theorem, Proc.Am. Math. Soc., 81, 89–92 (1981).
6. Buck, R.C., Bounded continuous functions on a locally compact space, Michigan Math.J., 5(1958).
7. Bucur, I., A version of Stone-Weierstrass theorem based on a Gavriil Paltineanu result in approximation theory, Proceedings of the Mathematical and Educcational Symposium of Department of Mathematics and Computer Science, 3–7, UTCB (2014).
8. Cristescu, R., Spatii lineare topologice, Ed.Academiei R.S.R., Bucuresti, 1974.
9. Cristescu, R., Topological vector spaces, Nordhoff International Publishing Leyden, The Netherlands, 1977.
10. Glicksberg, I., Measures orthogonal to algebras and sets of antysimmetry, Trans.Amer.Math.Soc., 105, (1962), 415–435.
11. Goullet de Rugy, A., Espaces de fonctions ponderables, Israel J.Math., 12 (1972), 147–160.
12. Jewett, R. I., A variation on the Stone-Weierstrass theorem, Proc.Amer. Math.Soc. 14(1963), 690–693.
13. Korovkin, P., P., Linear Operators and Approximation Theory, Moscow, Fizmatgiz, 1959.
14. Kravvaritis, D., Paltineanu, G., A density theorem for locally convex lattices, Abstract and Applied Analysis, 5 (2004) 387–393
15. Kravvaritis, D., Paltineanu, G., Some density theorem for complex locally convex lattices, Revue Math.Pure et Appl., 53 (2008), 2–3,167–179.
16. Lorentz, G, G., Approximation of Functions, HOLT, RINEHART AND WINSTON, 1966.
17. Luxemburg, W.A., Zaanen, A.C., The linear modulus of an order bounded linear transformation, I., Indag.Math. 33 (1971), 422–434.
18. Machado, S., On Bishop's generalization of the Stone- Weierstrass theorem, Indag.Math. 39,(1977), 218–224.
19. Nachbin, L., Elements of approximation theory, D.Van.Nostrand, Prinston, 1967.
20. Nachbin, L., Weigthed approximation for algebras and modules of continuous functions: real and self-adjoint complex cases, Ann. of Math., 81 (1965), 289–302.

© The Editor(s) (if applicable) and The Author(s), under exclusive licence
to Springer Nature Switzerland AG 2020
I. Bucur, G. Paltineanu, *Topics in Uniform Approximation of Continuous Functions*,
Frontiers in Mathematics, https://doi.org/10.1007/978-3-030-48412-5

21. Niculescu, C.,Paltineanu,G.,Vuza,D.,Interpolotion and approximation from the M-theory point of view, Revue Math.Pure et Appl., 6(1993), 531–544.

22. Pǎltineanu, G., Antialgebraic sets with respect to a subspace of , Revue Math.Pure et Appl., 10 (1976), 1391–1397.

23. Pǎltineanu, G., Teoreme de densitate in spatii de functii continue I, St.Cercet.Mat, 5(1978), 521–554.

24. Paltineanu, G., Teoreme de densitate in spatii de functii continue II, St.Cercet.Mat, 6(1978), 629–653.

25. Pǎltineanu, G., A generalization of the Stone-Weierstrass theorem for weighted spaces, Revue Math.Pure et Appl., 7 (1978), 1065–1068.

26. Pǎtineanu, G., Elemente de teoria aproximarii functiilor continue, Ed.Academiei R.S.R., Bucuresti, 1982.

27. Pǎtineanu, G., Vuza, D., Interpolating ideals in locally convex lattices, Revue Math.Pure et Appl., 41 (1996), 5–6, 379–396.

28. Pǎtineanu, G., Vuza, D., A generalization of the Bishop theorem for locally convex lattices of (AM)-type, Rendiconti del Circolo Matematico di Palermo, Serie II, Soppl. 52 (1998), 687–694.

29. Pǎtineanu,G., The complex locally convex lattices.Approximation lattice elements by using one of its own subspaces, Order Sructures in Functional Analysis, Editor Romulus Cristescu, Vol V, Ed Academiei Romane, 2006, 175–196.

30. Pǎtineanu, G., Some applications of Bernoulli's inequality in the approximation theory, Rom.J.Math.Comput.Sci. 4 (1), pp 1–11 (2014).

31. Pǎltineanu, G., Bucur, I., Some Density Theorem in the Set of Continuous Functions with Value in the Unit Interval, Mediterr, J. (2017) 14:44, published online March 2, 2017.

32. Pinkus, A., The Weierstass Approximation Theorems, Survey in Approximation Theory, Vol.1, 2005, pp. 1–37.

33. Prolla, J, B., Bishop's generalized Stone-Weierstrass theorem for weighted spaces, Math.Anal. 191 (1971), 283–289.

34. Prolla, J, B., On Von Neumann's variation of the Weierstrass-Stone theorem, Numer. Funct. Anal.and.Optimiz, 13 (3 and 4), 349–353 (1992)

35. Ransford, T.J., A short elementary proof of Bishop-Stone Weierstrass theorem, Math.Proc. Camb. Phil.Soc.96 (1984), 309–311.

36. Rudin, W., Functional Analysis, McGraw-Hill Book Company, 1973.

37. Rudin, W., Analizǎ realǎ si complexǎ, Editia a treia, Theta, Bucuresti, 1999.

38. Schaffer, H., Topological vector spaces, The Mc.Millmann Comp.New York, 1966.

39. De Schipper, A note on the modulus of an order bounded linear operator between complex vector lattices, Indag.Math. 41 (1979), 38–53.

40. Stone, M.H., The generalized Weierstrass approximation theorem, Math.Mag. 21 (1947–1948), 167–184 and 237–254 MR 10.p.255.

41. Summers, W.H., Dual spaces of weighted spaces, Trans.Amer.Math.Soc. 151, (1970) 323–333(1970)

42. Výborný, R., The Weierstrass theorem on polynomial approximation, Mathematica Bohemica, 130 (2005), No2, 161–165, 2004.

43. Weierstrass, K., Uber die analytische Darstellbarkeit sogennanter willkurlicher Functiononen reller Argumente, Sitzungsherichre der Acad.Berlin (1885), 633–639, 789–805.

44. Vuza, D.T., Elements of the theory of modules over ordered rings, Order Sructures in Functional Analysis , Vol 2, Editura Academiei Romane, Bucuresti, 1989, 175–283.

Index

Printed in the United States
By Bookmasters